Artificial Intelligence and Machine Learning for Smart Community

Intelligent systems are technologically advanced machines that perceive and respond to the world around them. *Artificial Intelligence and Machine Learning for Smart Community: Concepts and Applications* presents the evolution, challenges, and limitations of the application of machine learning and artificial intelligence to intelligent systems and smart communities.

- Covers the core and fundamental aspects of artificial intelligence, machine learning, and computational algorithms in smart intelligent systems

- Discusses the integration of artificial intelligence with machine learning using mathematical modeling

- Elaborates concepts like supervised and unsupervised learning, and machine learning algorithms, such as linear regression, logistic regression, random forest, and performance evaluation matrices

- Introduces modern algorithms such as convolutional neural networks and support vector machines

- Presents case studies on smart healthcare, smart traffic management, smart buildings, autonomous vehicles, smart education, modern community, and smart machines

Artificial Intelligence and Machine Learning for Smart Community: Concepts and Applications is primarily written for graduate students and academic researchers working in the fields of computer science and engineering, electrical engineering, and information technology. Seasonal Blurb: This reference text presents the most recent and advanced research on the application of artificial intelligence and machine learning on intelligent systems. It will discuss important topics such as business intelligence, reinforcement learning, supervised learning, and unsupervised learning in a comprehensive manner.

Artificial Intelligence and Machine Learning for Smart Community

Concepts and Applications

Edited by
T. V. Ramana, G. S. Pradeep Ghantasala,
R. Sathiyaraj, and Mudassir Khan

CRC Press
Taylor & Francis Group
Boca Raton London New York

CRC Press is an imprint of the
Taylor & Francis Group, an **informa** business

First edition published 2024
by CRC Press
2385 Executive Center Drive, Suite 320, Boca Raton, FL 33431

and by CRC Press
4 Park Square, Milton Park, Abingdon, Oxon, OX14 4RN

CRC Press is an imprint of Taylor & Francis Group, LLC

ISBN: 978-1-032-52606-5 (hbk)
ISBN: 978-1-032-52970-7 (pbk)
ISBN: 978-1-003-40950-2 (ebk)

DOI: 10.1201/9781003409502

Typeset in Minion
by Apex CoVantage, LLC

Contents

CHAPTER 3 ▪ A Detailed Case Study on Various Challenges in Vehicular Networks for Smart Traffic Control System Using Machine Learning Algorithms 51

BANDI VAMSI, BHANU PRAKASH DOPPALA, MOHAN MAHANTY, D. VEERAIAH, J. NAGESWARA RAO, AND B.V. SUBBA RAO

Preface

CONVENTIONAL HUMAN LIFE HAS BEEN MIGRATED TO THE DIGITAL realm thanks to modern technologies. It has been observed that many of the advanced applications have been developed with the integration of advanced smart technologies like artificial intelligence (AI), machine learning, big data analytics, business intelligence, and the Internet of Things. Among these technologies, AI drives industries to get automated and the ability to learn and apply optimal decisions at a given time. The rapid evolution of AI and machine learning (ML) has equipped machines with the ability to learn. AI has transformed most industry sectors like retail, manufacturing, finance, healthcare, and media and continues to invade new territories. ML will be an integral part of all AI systems, large or small. Connected AI systems will allow machine learning algorithms to "continuously learn" from real data on the internet. ML will assist machines in better understanding information's meaning and intent. AI and ML have also made good advances in education and learning, as well as smart healthcare. During the pandemic, this technology aided in a wide range of applications. Many researchers are working across the globe to design and shape the future of the smart world in a pioneering way.

This book focuses on cutting-edge research in the area of smart applications of AI and ML, ranging from the fundamentals/concepts, novel algorithms, mathematical models, case studies, and so on. The primary objective of this book is to recognize the trends in the field of AI integrated with ML and to dig out various efficient solutions for promoting smart applications for a better society. This book investigates the recent technological advancements and achievements in the said area and aims to discover the recent trends in AI and ML to promote and publish original articles in the relevant field of integration of ML and AI in various aspects of smart applications. It will also provide research directions for AI and ML in terms of offering better solutions for a smart and safe society.

The book provides the fundamental concepts and techniques with case studies employing AI and ML for smart applications. Students across various domains, researchers, academicians, and industrialists will have an opportunity to understand the notions of AI and ML and analyze and design an innovative application for the smart community.

About the Editors

T. V. Ramana is Professor of Computer Science and Engineering, JAIN University, Bangalore, India, where he heads the Cloud Technology and Information Security Group. He received his M.Tech. and Ph.D. from the Jawaharlal Nehru Technological University (JNTU), Hyderabad. His research interests include software engineering, computer system architecture, machine learning, and the Internet of Things (IoT). He is also Associate Dean of the Faculty of Engineering and Technology. He has researched extensively on cloud computing and computer system architecture. His current research focuses primarily on the IoT and machine learning.

G. S. Pradeep Ghantasala is currently a professor in the Department of Computer Science and Engineering, Alliance University, Bangalore, India. He has published various papers in reputed journals such as *SCI*, Scopus-indexed journals, UGC journals, conferences, books, book chapters, and national and international patents. He is also the editor and reviewer of various journals. His research interests include machine learning, deep learning, healthcare applications, and software engineering applications. Programming, machine learning, and artificial intelligence are just a few of the courses he has taught for both UG and PG students. He has published more than 30 manuscripts in scientific publications.

R. Sathiyaraj is currently working as an assistant professor in the Department of CSE, GITAM University, Bangalore. He received his B.Tech. in Information Technology from Anna University, Chennai, India, in 2009; M.Tech. from Vel Tech University, Chennai, India, in 2012; and Ph.D. in Computer Science and Engineering from Anna University, Chennai, India, in 2020. He has contributed to two books and is the lead editor of two books on various technologies. He has to his credit five patents and more than 20 articles and papers in various refereed journals and international conferences. His research interests include ML, big data analytics, and intelligent systems.

Mudassir Khan is currently working as an assistant professor and former Head in the Department of Computer Science at College of Science & Arts Tanumah, King Khalid University Abha, Saudi Arabia. He has more than 10 years of teaching experience at the King Khalid University of Saudi Arabia. He has published more than 25 papers in international journals (*SCIE, ESCI, Scopus, UGC Care*), presented in conferences of the IEEE and Springer Nature, and has to his credit one patent. He is a member of various technical/professional societies such as IEEE, UACEE, Internet Society, IAENG, and CSTA. His research interests include big data, IoT, deep learning, machine learning, eLearning, fuzzy logic, image processing, cyber security, and cloud computing.

Contributors

Ambati Renuka
Graduate Scholar
Faculty of Engineering and
  Technology
Sri Ramachandra Institute of
  Higher Education and Research
Chennai, Tamil Nadu, India

Ayushmaan Das
Graduate Scholar
Faculty of Engineering and
  Technology
Sri Ramachandra Institute of
  Higher Education and Research
Chennai, Tamil Nadu, India

B.V. Subba Rao
Department of Information
  Technology
PVP Siddhartha Institute of
  Technology
Andhra Pradesh, India

Bandi Vamsi
Department of Artificial
  Intelligence & Data Science
Madanapalle Institute of
  Technology & Science
Madanapalle – 517326, India

Bhanu Prakash Doppala
Data Analytics
Generation Australia
Level 35, 88 Phillip St, Sydney
  NSW 2000, Australia

D. Veeraiah
Department of Computer
  Science and Engineering
Lakireddy Bali Reddy
  College of Engineering
  (Autonomous)
Mylavaram, Andhra Pradesh
  521230, India

E. M. Roopa Devi
Associate Professor
Department of Information
  Technology
Kongu Engineering College
Erode, India

Heena Wadhwa
Chitkara University
  Institute of
  Engineering &
  Technology
Chitkara University
Punjab, India

Iyapparaja M.
School of Information Technology
and Engineering
Vellore Institute of Technology
Vellore, Tamilnadu, India

J. Nageswara Rao
Department of Computer Science
and Engineering
Lakireddy Bali Reddy
College of Engineering
(Autonomous)
Mylavaram, Andhra Pradesh,
India

Jayanthi Ganapathy
Assistant Professor
Faculty of Engineering and
Technology
Sri Ramachandra Institute of
Higher Education
and Research
Chennai, Tamil Nadu, India

Mohan Mahanty
Department of Computer
Science and Engineering
Vignan's Institute
of Information
Technology (A)
Visakhapatnam, India

Ochin Sharma
Chitkara University Institute
of Engineering &
Technology
Chitkara University
Punjab, India

Ojas Sharma
Chitkara University Institute of
Engineering and Technology
Chitkara University
Punjab, India

Pooja Sharma
Assistant Professor, Department of
Computer Science & Engineering
IKG Punjab Technical University
Kapurthala, Punjab, India

Raj Gaurang Tiwari
Chitkara University Institute of
Engineering & Technology
Chitkara University
Punjab, India

S. Anitha
Assistant Professor (Sr.G),
Department of Information
Technology
Kongu Engineering College
Erode, India

Tejinder Kaur
Chitkara University Institute of
Engineering and Technology
Chitkara University
Punjab, India

Thanushram Sureshkumar
Graduate Scholar
Faculty of Engineering and
Technology
Sri Ramachandra Institute of
Higher Education and Research
Chennai, Tamil Nadu, India

Vamsi Krishna
Graduate Scholar
Faculty of Engineering and
    Technology
Sri Ramachandra Institute of
    Higher Education and Research
Chennai, Tamil Nadu, India

Venkatasaichandrakanth P.
School of Information
    Technology and
    Engineering
Vellore Institute of
    Technology
Vellore, Tamilnadu, India

A Detailed Study on Deep Learning versus Machine Learning Approaches for Pest Classification in Field Crops

Venkatasaichandrakanth P. and Iyapparaja M.

1.1 INTRODUCTION

1.1.1 Background

Globally, agriculture supplies food for both people and animals; hence, the main goal must be to increase crop yields by controlling pests [1]. Climate change, a severe lack of arable land and water supplies, weeds, disease, and pests are some of the important risks to agriculture [2]. The words "pest" and "peste," both of which mean "plague" or "contagious illness," are derived from the Latin word *pestis* and the French word *peste* [3]. Agricultural crops have been continuously threatened by pests and changes in the environment [4]. Monitoring and evaluating pests and insect control are essential for ensuring crop quality and safety [5]. It takes a lot of time, money, and labor to identify and control pests. To control pests, pesticides are used, but using them excessively can harm both plants and people [6]. The creation of an automatic method for spotting pests in crops is essential for resolving the aforementioned problems. Photos of the recognized plants are acquired, and these images are then analyzed using image-processing techniques to ascertain the types of pests and illnesses. Using extracted features, automated classification and identification tasks are carried out. Artificial intelligence (AI) techniques are used to extract these properties from photos. As a result, the detection of pests and diseases might be totally automated and finished more quickly [7].

Machine-vision approaches are effective and labor-saving when compared to manual identification. In the area of machine vision, the detection of plant pests and diseases is a main area of study. Image processing is a method that acquires images of plants to detect diseases and pests [8]. Initial applications of machine vision-based disease and pest detection have occurred in agriculture, and they have partially replaced traditional naked-eye identification. Computer vision-based methods for pest detection are based on traditional image processing. It is common to use algorithms and classifiers with features designed by hand [9]. The imaging method for this technology is determined based on the various features of pests and plant diseases, and the light source and camera angle are chosen accordingly. This creates uniform illumination across images. The well-designed image reduces the difficulty of classical algorithms, but the application cost may increase. However, in the natural environment, it is very difficult to build a classical algorithm that eliminates scene changes on recognition results [10]. Table 1.1 presents the abbreviations.

1.1.2 Dataset

The type of dataset used is the image dataset, and it is collected from the Kaggle website. It consists of nine types of pest classes, which are aphids, beetles, grasshoppers, mosquitos, stem borer, mites, bollworms, armyworms, and sawflies. Three hundred images from the training dataset per each pest class and 50 images from the test dataset per each pest class were utilized [11].

1.1.3 Machine Learning

In the AI domain, machine learning (ML) is identified as an important area. The human mind learns from past data and experiences and makes decisions in the future based on that knowledge. Any traditional computer algorithm operates according to the developer's instructions [12]. ML algorithms consist of different types. They include supervised learning, which is a classification algorithm based on some experience that predicts

TABLE 1.1   Abbreviations

ANN	Artificial Neural Network
CNN	Convolutional Neural Network
DCNN	Deep Convolutional Neural Network
DL	Deep Learning
DRSF	Domain-Related Specific Features
FCL	Fully Connected Layer
F-RCNN	Fast Region-based Convolutional Neural Network
FT	Fine Tuning
GIST	Global Image descriptor
HOG	Histogram of Oriented Gradients
KNN	K-Nearest Neighbors
LRN	Local Response Normalization
MAE	Mean Absolute Error
MAP	Mean Average Precision
ML	Machine Learning
MLP	Multi-layer Perceptron
NB	Naive Bayes
NPA	Negative Predictive Accuracy
PPA	Positive Predictive Accuracy
RCNN	Region-based Convolutional Neural Network
ReLu	Rectified Linear activation Unit
RESNET	Residual Neural Network
RF	Random Forest
SVM	Support Vector Machine
TL	Transfer Learning
UAV	Unmanned Aerial Vehicle
VGG	Visual Geometry Group
YOLO	You Only Look at Once

the output; unsupervised learning, which is a clustering algorithm based on some statistical and similarity-based methods that predict the output; reinforcement learning, which is a method that predicts the output based on learning from mistakes; and finally semi-supervised learning, which is a combination of classification and clustering. Normally, supervised learning consists of labels. In clustering we have only un-labels; by using some statistical methods, it finds the labels for un-labels. However, Semi-supervised learning is a technique that combines a limited number of labeled labels with a larger number of unlabeled labels. It leverages the labeled data to infer or predict labels for the unlabeled data. Recently, combining image processing and ML has been a popular topic of study and

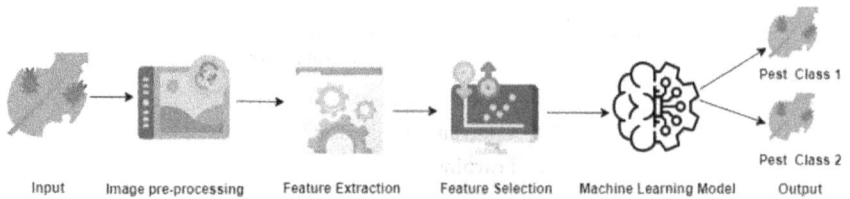

FIGURE 1.1   Image processing and machine learning classification workflow.

application for researchers [6]. Many features may be extracted from the collected image using machine vision algorithms, which can be utilized to distinguish the objects [13]. The extracted features are very important as they contain information about the image. This information is given to the machine learning algorithms, which classify and detect the pests in the different field crops. Machine learning depends on feature selection for classification; when the best features are identified to be passed on to the ML classifier, it increases the complexity [14]. Figure 1.1 shows the procedure for pest recognition from captured images. After capturing an image, its quality is improved by using image pre-processing techniques. A computer reads the image as a matrix of numbers. The matrix size is based on the number of pixels. Pixels are also called a feature.

1.1.3.1 Image Feature Extraction

Extraction of features is defined as taking out relevant pixels from image segmentation, and it is also known as dimensionality reduction. It is used for shrinking the size of a feature space without removing information from the given feature space [15,16], which helps to make an accurate classification of the anomaly. The features are classified into two types, which are domain-related specific features (DRSF) and general or normal features. DRSF are fully dependent features such as conceptual features and human faces, whereas general features are independent features related to application, such as color, texture, and shape. Based on some analysis, these are further classified into pixel-level features, which are evaluated at each pixel, such as location and color. Local features are defined as image patches, which are also known as a small group of pixels such as image segmentation, whereas global features are defined as an entire image [17,18]. Image feature extraction methods are mentioned below.

1.1.3.1.1 Color Feature Extraction

In image retrieval, color feature extraction is one of the most important visual features. It consists of methods such as histogram intensity, color histogram, and Zernike chromaticity distribution moments [19,20].

1.1.3.1.2 Texture Feature Extraction

Texture feature defines the physical arrangement of space. It consists of methods such as edge detection and grey-level co-occurrence matrix [19].

1.1.3.1.3 Shape Feature Extraction

It is a concept in feature extraction and is used for recognition. It is very difficult to recognize correctly without shape feature. It consists of methods such as vertical segmentation, horizontal segmentation, and binary image algorithm [19].

1.1.3.2 Image Feature Selection

After extracting the features from the image, these feature vectors are correlated using feature selection. It is used to decrease the number of variables using data dimensionality reduction methods [21,22]. Image feature selection methods are mentioned below.

Independent Component Analysis: It is a transformation approach used for decomposing the data into subparts.

Principal Components Analysis: It is transformation approach used for identifying information in the data and focusing on their similarities and differences.

Linear Discriminant Analysis: It is a general method for performing dimensionality reduction [21].

1.1.3.3 Machine Learning Techniques

After feature selection, to classify the pest ML, classification techniques are used. Some of the ML techniques are as discussed below.

1.1.3.3.1 Support Vector Machine

Support vector machine (SVM) is used in classification- and regression-related mechanisms in ML. The main objective of SVM is to find the decision margin that can differentiate the n-dimensional feature space into a number of classes. Here, hyperplane is the best decision margin. To create a hyperplane, it helps to select the extreme points, which are called support vectors, and hence the algorithm is called as SVM [23]. Figure 1.2 shows a block diagram of SVM. In order to achieve classification, an SVM seeks out the hyperplane that maximizes the margin between the two classes. It is necessary to maximize the width of the margin in order to establish an ideal hyperplane (z).

$$\vec{z}.\vec{y}+c=0 \tag{1.1}$$

In Equation 1.1 belongs to margin equation. Based on this margin equation value, we can find the item class is defined as belonging to class 1 or class 2.

$$(\vec{z}.\vec{y}+c)\geq1 \tag{1.2}$$

In Equation 1.2 belongs to class 1.

$$\vec{z}.\vec{y}+c\leq-1 \tag{1.3}$$

In Equation 1.3 belongs to class 2.

For example, develop an SVM model that can classify between a tiger and lion present in an image. Let's imagine a strange tiger that looks like a lion. Use the SVM method to create such a model. First, train our model

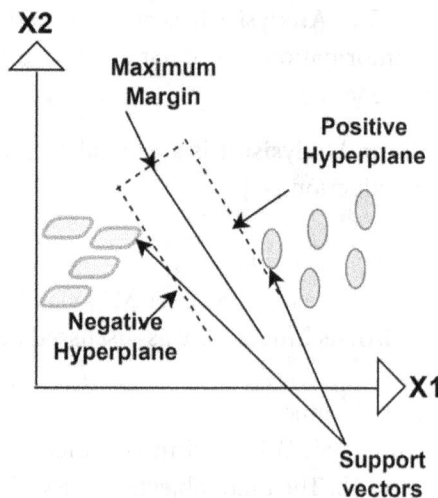

FIGURE 1.2   Workflow for support vector machine.

with multiple pictures of tigers and lions so that it closely matches the variety of features of tigers and lions before validating it with this unknown image of a tiger or lion. As a result, when it constructs a boundary line between two sets of data of tiger and lion, it will focus on the most concerning situations involving tigers and lions. The support vectors will be used to categorize it as a tiger or lion based on the input image. Based on some conditions, SVM can be divided into two categories. When a dataset can be divided into two classes by a single straight line, it is said to be linearly separable, and the used classifier is known as a linear SVM classifier. A dataset is considered non-linearly separated when it cannot be divided into two groups using a straight line, and the classifier used is known as a non-linear SVM classifier [23].

K-Nearest Neighbor Classifier: K-nearest neighbor (KNN) classifier is the simplest ML algorithm; it works on the principle of supervised learning. It is known as a lazy learner strategy since the training dataset is kept rather than immediately used for learning. Instead, it performs an action while classifying data using the dataset. When new data is received, the KNN approach simply preserves the information from the training phase and categorizes it into relatively similar categories to the new data [24].

Consider the following scenario. There is a picture of an animal that resembles both a tiger and a lion, but not sure which one is present in the figure. KNN algorithm, which relies on a similarity metric, is thus appropriate for this issue. Based on the fact that the features in the new dataset that are most equivalent to those in the lion and tiger images, this model will identify the data as either belonging to the lion class or the tiger class. Figure 1.3

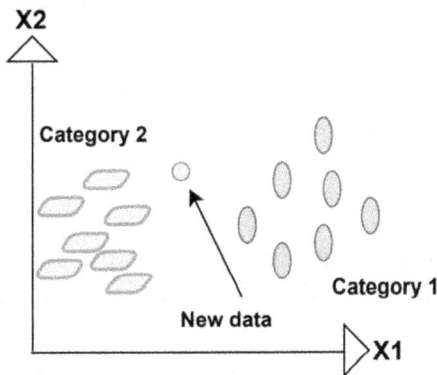

FIGURE 1.3   Block diagram of K-NN.

describes the block diagram of the K-NN classifier. Selecting the K neighbor is the first stage. Calculating the Euclidean distance between k neighbors is the next step. Equation 4 describes the Euclidean distance between the two points E1 (m1, n1) and F1 (m2, n2) as

$$\text{Euclidean Distance} = \sqrt{(m2 - m1)2 + (n2 - n1)2} \qquad (1.4)$$

The next step is to choose the K neighbors based on the distance estimate. The following step is to count how many of these k neighbors' data points fall into each category and, finally, to add the extra data points to the category with the highest neighbor count [24].

1.1.3.3.2 Decision Tree

Although it can be applied to situations involving classification and regression, for classification-related issues, the supervised learning method named decision tree (DT) is frequently used. It resembles a tree-structured classifier in which the features of a dataset are represented by the inner tree nodes and leaf tree nodes, which indicate the classification output. Basically, DTs consist of DT nodes and leaf tree nodes that indicate the issue or decision based on specified conditions; on the other hand, they display the results of those decisions and do not have any more branches [25].

Figure 1.4 shows the process of DT. To predict the class of the provided dataset, the algorithm starts at the root node and works its way up. This approach goes to the subsequent node by contrasting the values of the record variable and the root node property. The biggest challenge while designing a DT is deciding which attribute is ideal for the root node and sub-nodes. There is a method known as attribute selection measure that can be used to solve such challenges. To do this, it consists of two techniques:

1. Information Gain

2. Gini Index

The algorithm confirms the attributes with the other leaf nodes before going on to the next node. This keeps going until it encounters the tree's leaf node [25].

Random Forest: It is an ensemble learning algorithm that combines DTs on various subsets of the dataset, and the final results are averaged to improve the predictability of the dataset [26].

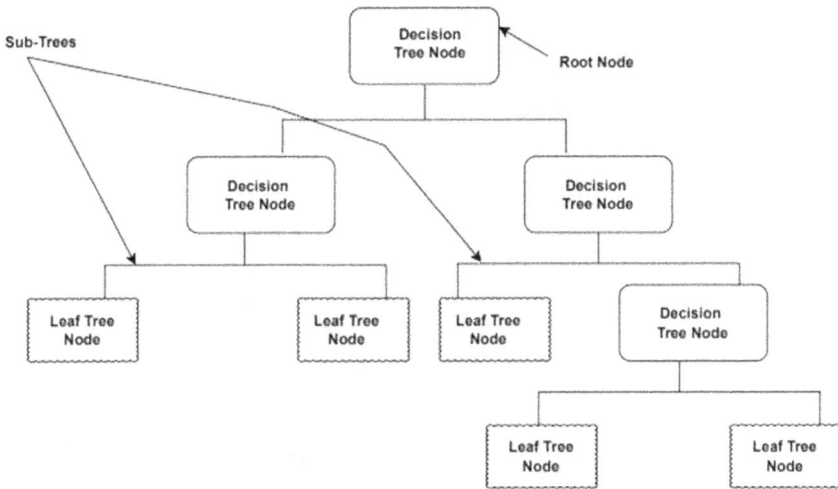

FIGURE 1.4   Workflow process of decision tree.

Uses of random forest: A model can be trained faster than with other ML algorithms. Next, with the enormous dataset, it operates effectively and makes output predictions with effective accuracy, and accuracy can be preserved even when a major piece of the data is missing.

Working procedure of random forest: The first step is to choose the *n* number of subsets from the training or labeled data. The second step is to construct the DT from subsets of training data. The third step is to select N number of DTs that you want to construct. The fourth step is to repeat the second and third steps. The final step is to locate each DT's estimates for any new data items and then group them into the category with the most support. Figure 1.5 shows the workflow process of random forest [26].

A lot of limitations exist with traditional machine learning algorithms for image recognition. One of the key drawbacks of this method is that feature engineering by hand is time-consuming, and the parameters must be adjusted based on the sorts of prediction challenges.

1.1.4 Deep Learning

The branches of AI are ML and DL. Since 2016, DL techniques have been successfully used in different research areas, including image processing and speech recognition [27]. DL has improved in its ability to solve complex tasks during the past few years. The growth of effective autonomous

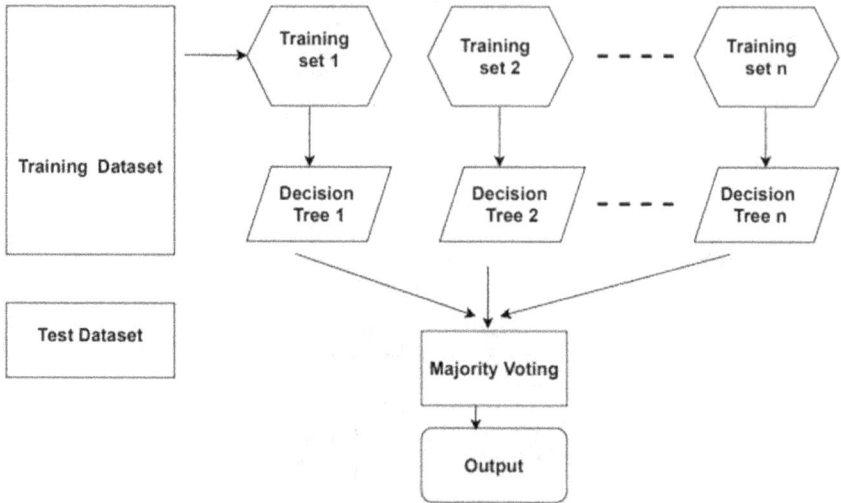

FIGURE 1.5 Workflow process of random forest.

systems has been fueled by the rising demand for this type of technology, especially given its potential use in farming [10]. DL-based automatic feature extraction solves the limitations of handcrafted features and complex issues like the classification of images. In the area of visual objects and ML, DL systems such as neural networks have made significant development [7]. Deep learning, according to a recent study, takes the use of the extraction of features in the images, through the adaptive learning of artificial neurons; this is implemented and no longer impacts the process of artificially identifying relevant features [28]. Feature extraction is done in a somewhat automatic manner throughout DL algorithms. Researchers are encouraged to extract features with the minimum amount of human work and field knowledge possible [29]. Large numbers of samples are necessary for ML and DL algorithms, and they detect the features of the inputs and predict the outputs using decision rules. DL models based on Convolutional Neural Networks (CNNs) are used to solve different problems in the agriculture domain, including classification of fruits, leaf disease detection, pest detection, and weed identification [14]. The CNN is an efficient method of analyzing visual imagery. A neural network consists of three layers, such as the input layer, which gives images as input; hidden layer, which does some operations like convolutional, pooling, normalizing, and FC layers; and output layers does operations like classifications. The CNN consists of a number of trainable hyperparameters (weights)

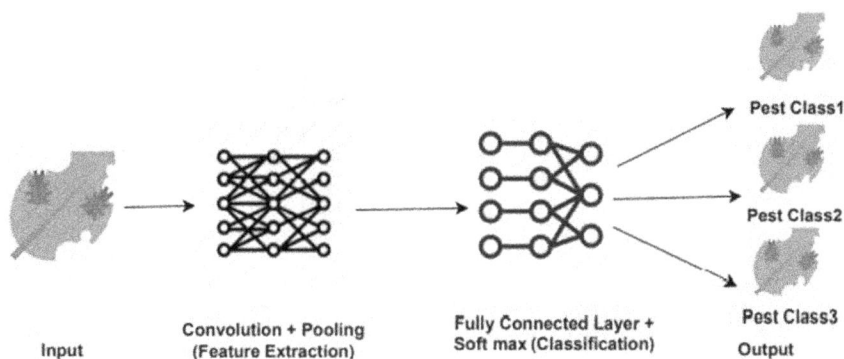

Input — Convolution + Pooling (Feature Extraction) — Fully Connected Layer + Soft max (Classification) — Pest Class1, Pest Class2, Pest Class3, Output

FIGURE 1.6   Primitive workflow of the convolution neural network.

that allow it to learn the sequential relationship between the inputs and perform a detection or classification task [30].

Before deep learning, most of the aforementioned machine learning approaches relied completely on hand-crafted features. This resulted in the process of constructing pipelines, which is highly time-consuming since images need to be captured and aggregated in carefully controlled conditions. As specified in Figure 1.6, the CNN [31] is significant for extracting features and learning from training data or labeled data during the process of training. The CNN computes the activation map by convoluting the kernel on the image. It further utilizes the pooling layer, which takes the activation map as input with the objective of decreasing the dimension of the activation map. Thus, any change in the input image, even a small translation or rotation has the possibility of being absorbed. Furthermore, the convolution and max pooling layers are used repeatedly for extracting features from the activation map. Moreover, the convolution layers are the most significant components of CNNs. As input, the obtained features are given to the fully connected layer (FCL), and finally, the output is determined as the probability of every class. A network topology with units for the image and the various classes is present in the input and output layers in this situation. Various actions associated with CNN include loss function, activation function, regularization, decreasing the number of parameters and development in design [32]. Some of the deep CNN's algorithms are discussed next.

1.1.4.1 AlexNet

AlexNet was discovered by Alex Krizhevsky and is a CNN. It is the first and the oldest6 CNN model that was employed to increase the model's

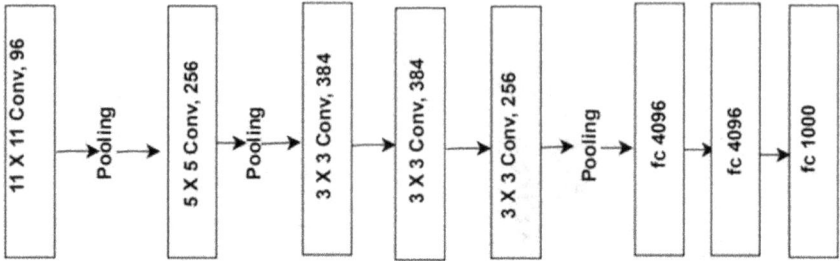

FIGURE 1.7   AlexNet architecture.

efficiency. It has a total of five convolutions, three max pools, two fully connected layers, and two normalized layers. Here each convolution layer has a kernel and applies non-linear functions like a ReLu and Leaky ReLu. The max pools are used to conduct max-pooling; here, max-pooling means that after performing filter operations on the input image on convolution, it generates a feature map from that selects maximum number of localities [33,34]. Figure 1.7 shows the architecture of AlexNet.

1.1.4.2 ZFNet

ZFNet is a variant of AlexNet. The fact that ZF Net utilized 7x7-sized filters while AlexNet used 11x11-sized filters was a key distinction between the two methods. This is predicated on the notion that by using bigger filters, you lose more pixels that you could have saved by using smaller filters in the earlier convolution layers. As we look further, there are more filters. Additionally activated using ReLus, this network underwent batch stochastic gradient descent training [35].

1.1.4.3 VGGNet

The VGG network was first invented by Andrew Zisserman and Karen Simonyan in 2013, and their exact model was released for the ImageNet Challenge in the year 2014. In honor of their affiliation with the Visual Geometry Group Department at the University of Oxford, they gave it the name VGG [36].

1.1.4.4 VGG16 Architecture

Input to the convolutional 1 layer is an RGB-colored image with a fixed size of 224 by 224. Filters were used with a 3*3 receptive field, which was relatively small. After that, the input image is processed through the convolutional (conv.) layer stack. One of the ideas uses 11 convolutional filters from VGG, which may be thought of as a linear variation in the input fields. The vertical padding of the convolved input is such that the pixel size is kept after convolution while using 3*3 convolution layers and a fixed convolution stride of 1 pixel [36,34]. The pooling is done via five maximum pooling layers, some of which follow the convolution layers. Max-pooling is performed using Stride 2 over a 2*2 pixel window. A stack of convolutional layers is followed by three FCLs (which differs in depth depending on the architecture). The FCLs are constructed in a similar fashion across all networks. Rectification (ReLu) non-linearity unit is a feature shared by all hidden layers. Notably, all but one of the networks does not employ local response normalization (LRN), which expands processing time and uses more memory without enhancing performance on the ILSVRC dataset [36]. Figure 1.8 shows the architecture of the VGG16 network.

1.2 RELATED WORK

1.2.1 Pest Classification in Field Crops Based on ML Techniques

Lorris Nanni et al. [37] developed a classification model for crop insect pests. Images of the Deng and IP102 insect pests were acquired online for this purpose. Assessing the effectiveness of their ensemble as well as the performance of each pre-processing on the smaller dataset, our top ensembles perform at the level of research specialists (92.43%), whereas on the larger dataset, they achieve an accuracy of 61.93%. Thenmozhi Kasinathan et al. [38], the authors of this study, used sample butterfly photographs that were downloaded from the internet together with existing or publicly available datasets named Xie and Wang. By applying image pre-processing techniques that take into account a variety of characteristics like texture, shape, color, histogram of oriented gradients (HOG), and global image descriptor (GIST), they give special focus to the

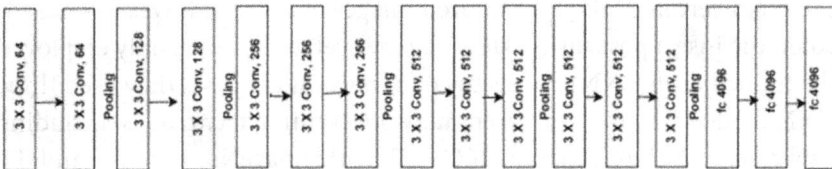

FIGURE 1.8 VGG16 network architecture.

classification of field insect pests. The taxonomy of insect pests included each of these characteristics. SVM, RF, and other classifiers make up the base classifier. Ensemble classifiers also incorporate XGBoost. The outcomes demonstrated that ensemble classifiers outperform base classifiers. Miranda J. et al. [39] built a system for automatic pest identification and extraction in paddy crops. The authors captured the images on the agricultural land using a camera. The collected photos are converted from RGB to grayscale. The authors employed background modeling to find the pests in the photos and then a median filter to get rid of noise from various lighting situations. It is easy to identify objects in an image. Each coordinate was calculated and preserved by scanning the image both horizontally and vertically. This paper's encouraging findings showed that both material and procedure improvements needed to be addressed. As a result of the above, the model accuracy of the pest classification is fully dependent on a number of factors that are related to the author's choice. The choices involve which pre-processing techniques to use, which color spaces to use, for example RGB, grayscale, and so on, which relevant features to extract from the image, and which classifier to implement.

Table 1.2 summarizes recent works in machine learning–based pest identification research and highlights the image pre-processing methods, features, and classifiers employed, as well as the efficiency of the models developed in the various researches.

1.2.2 Pest Classification in Field Crops Based on DL Algorithms

Xi Cheng et al. [44] introduced a DCNN-based pest identification method. Deep residual learning is used in this identification method. The accuracy of AlexNet, ResNet, Residual networks 50, and 101 DCNN was taken in this work and compared to that of SVM and backpropagation neural network on the Xie dataset of 10 classes of pest photos. With the Xie dataset, the accuracy was first calculated using SVM, BP NN, and AlexNet. The accuracy of the AlexNet got better results. Using a UAV camera, Everton et al. [45] captured images of soybean leaves in order to count the insect pests using DCNN. However, the authors only employed the trained deep CNN models to count one kind of pest. They used three CNN models and three different network training techniques, including data augmentation, dropout, 100% FT with ImageNet weights, and TL with ImageNet weights, to reduce over-fitting and improve model performance. The results showed that, in terms of classification accuracy,

TABLE 1.2   Summary of Individual Articles of Machine Learning–Based Pest Classification

Article	Year	Dataset	Pre-processing technique	Features used	Models used	Performance
[40]	2022	Predefined Datasets (NBAIR, TNAU) and web sources	Image resize, image transformation – RGB to Lab color space, Image segmentation – Kmeans	Texture	SVM, KNN, and decision tree (DT) classifier	Accuracy
[41]	2021	Predefined Datasets	Data Annitation, Dataset Split, Image resize, Data augmentation	–	–	–
[42]	2022	Remote sensing data taken from USGS Earth Explorer	Spectral bands, Vegetation indices	Texture	SVM, ANN, RF	Accuracy
[7]	2021	Images were collected from chilli farm located at Malaysia	Segmentation techniques used to differentiate the back ground and foreground components of chilli leaves	Traditional and DL Feature based method	RF, SVM, ANN	Precision, Recall, F1-score
[38]	2021	Insect images collected from pre-defined datasets Xie and Wang	–	Texture, Color, Shape Features	bagging, RF, adaboost, KNN, MLP, SVM, NB	Accuracy
[43]	2019	–	Segmentation method is K-means clustering method	Texture Features	SVM	Accuracy, Sensitivity, Specificity, PPA, NPA

training DL models with FT outperformed other training techniques. Muammer Turkoglu et al. [46] implemented a system for the detection of pests and diseases. Images of pests and plant diseases gathered in the field are used to train this model. The authors analyzed the system in three parts. First, they used ensemble averaging to FT the CNN models using the TL approach. Next, a supervised classifier is trained employing SVM based on features generated from several CNN models using the early fusion method. Next, following that, several CNN models are used to feature extraction for the majority voting method through the use of SVM classifiers. Finally, all class labels are evaluated by a majority vote, with the system's final result being the most predicted label. The results showed that the proposed plant disease network majority voting model obtained better accuracy (97.56%) as compared with the remaining two methods. Madhuri Devi Chodey et al. [47] have introduced a novel hybrid deep learning model that is capable of identifying pests in a variety of plant species. In order to coordinate deep learning techniques, the authors devised a four-stage paradigm. Evaluating performance tests are run on a variety of publicly accessible benchmark datasets that were obtained from various types of agricultural land. The structural similarity index is used to calculate the MAE and AP (SSIM). These numbers are 0.99, 0.2, and 89.67%, respectively. According to an evaluation, the method's qualitative performance has been found to be capable of real-time monitoring and detection. The majority of the CNN architectures employed in this study's DL models for pest classification included VGG and AlexNet. These models were created and enhanced for use with the ImageNet database. These CNN architectures must be quite complex and have a large number of parameters that may be learned in order to categorize the input data taken from the ImageNet dataset. Additionally, since pest classification datasets are not as diverse as the ImageNet dataset, a substantially lighter CNN model is required.

Tables 1.3 and 1.4 summarize recent works in deep learning–based pest classification. They highlight various datasets used in each study as well as the evaluation of the DCNN models established in the various research studies. Accuracy, recall, F1-score, and precision are common model performance metrics. When DL models were compared to ML algorithms in this study, DL models outperformed ML methods.

TABLE 1.3 Summary of Deep Learning–Based Pest Classification Schemes

Article	Year	Crop	Objective	Dataset	Models used	Performance
[48]	2022	–	Pest detection	Deng and IP102 predefined datasets	ResNet50, GoogleNet, ShuffleNet, MobileNetv2, and DenseNet201	Accuracy
[49]	2022	Soybean	Pest detection and Classification	Images were collected from laboratory	YOLOv4	Accuracy, Precision, Recall, F1-score
[50]	2022	Wheat	pest Classification	Images were collected from wheat filed	Proposed mAlexNet and hybrid model (mAlexNet + BiLSTM)	Accuracy, Precision, Recall, F1-score, Specificity
[51]	2020	–	pest Classification	Images were collected by using camera and from internet	MatConvNet is a tool box available in MATLAB®	Accuracy
[52]	2021	Rice	Pest detection	NBAIR predefined dataset	AlexNet, GoogleNet, VGG 16	Accuracy
[53]	2021	–	Pest monitoring	16 pests categories were collected from field by using camera	CNN	mAP, Accuracy
[54]	2020	Soybean field	Pest detection and classification	Pest images were collected from soyabean leaves by using camera mounted in the UAV	Resenet-50, Xception, Inception-v3, VGG-16,19	Accuracy
[55]	2021	Citrus	Pest classification and recognition	Citrus images collected from field at Iran	Alexnet, VGG16, Resnet-50, Inception resnetV2, Proposed ensemble model	Accuracy, Recall, F1- measure
[56]	2021	Wheat	Recognition and Counting of wheat mite	Images were captured by camera in field	ZFnet, VGG-16	mAP, Accuracy
[57]	2020	Cotton	Pest classification	Images were captured by camera in field	Modified Resnet34	Accuracy

TABLE 1.4 Summary of Deep Learning–Based Pest Classification Schemes

Article	Year	Crop	Objective	Dataset	Models used	Performance
[58]	2020	-	Pest recognition	Images were collected from common agricultural and forestry pests	CPAFNET new deep CNN model	Accuracy
[59]	2020	-	Pest recognition	Images were collected from Internet	VGG-16, VGG19, ResNet50, ResNet152, GoogLeNet	Accuracy
[60]	2017	-	Pest detection and classification	Videos are collected from farmlands	Faster-RCNN with RPN, VGG-16, Alexnet	Accuracy
[61]	2017	-	Pest Classification	Own image dataset collected	Lenet-5, Alexnet	Accuracy
[62]	2021	-	Disease and pest classification	-	DCNN-Google, YOLO-V4	Accuracy
[63]	2021	Tomato	Pest and disease detection	Large scale images were collected in field by agriculture internet of things	YOLO-V3	F1- measure, Precision, Recall

1.3 OPEN-SOURCE TOOLS FOR DEEP LEARNING FRAMEWORKS

To speed up the task and achieve efficient results, various frameworks and libraries were used. By using these frameworks, training procedures have become easier as a result of their ease of use and their ability to modify the parameters of neural networks. Famous DL frameworks are Tensor Flow, Keras, Caffe, Pytorch, and so on.

Tensor Flow: Tensor flow was discovered by the Google Brainteam. It is a complete open-source machine learning platform [64].

Keras: F. Chollet designed Keras. An open-source software package, Keras offers a Python framework for deep neural networks. It also provides the Tensor Flow library interface [65].

Caffe: Learning Center and Berkeley Vision are known for creating Caffe. A deep neural framework called Caffe has an emphasis on modularity, performance, and expression [66].

Pytorch: A GPU-accelerated machine learning package with a focus on automatic differentiation and tensor computations is called Pytorch. In order to compete with Keras and Tensor Flow for the title of "most utilized," deep learning package Pytorch has become one of the most well-known deep learning libraries [67].

1.4 CHALLENGES IN THE AREA OF PEST CLASSIFICATION USING ML/DL

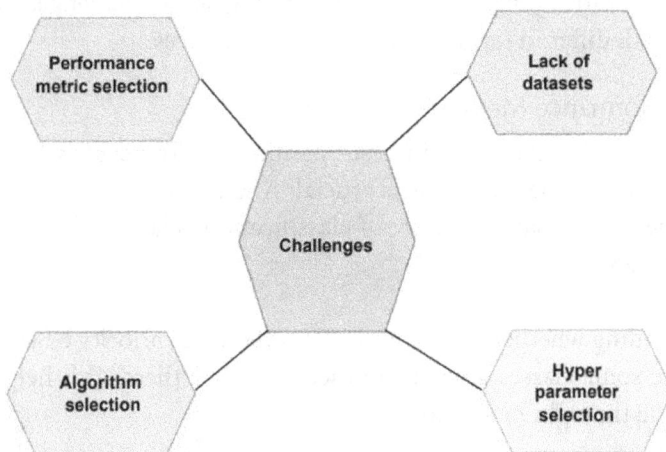

FIGURE 1.9 Challenges.

1.4.1 Lack of Dataset

To solve this problem, some existing methods shown here are

1. Open-source datasets

2. Transfer learning

3. Data augmentation or image augmentation

4. Synthesis data creation

1.4.2 Selection of Hyperparameters

Hyperparameters impact the CNN model's accuracy. Optimization techniques are used to tune hyperparameters, but deciding which technique to use for which application is very difficult. Selecting suitable hyperparameters for deep Convolutional Neural Network (CNN) models can pose a formidable challenge due to the intricate nature of these models. Determining which hyperparameters are optimal for a given deep CNN model is a complex task. Here are some common hyperparameters are learning rate, batch size, number of layers and units, kernel size etc.

1.4.3 Algorithm Selection

In certain cases, choosing a model comes down to knowing how the model handles various dataset sizes. It's also difficult to choose an algorithm. For any problem in machine learning or deep learning, so many methods can be applied and different models can be generated. The number of data rows and features is vital to consider while choosing an algorithm. The models will operate with a range of datasets depending on the challenge, and these datasets will differ in terms of data rows and features.

1.4.4 Performance Metric Selection

This is one of the challenges in pest classification. To decide which model to use, the performance metric is crucial. A selection criterion for classification metrics is that the choice of classification is largely determined by two variables.

1. Deciding whether to give every class the same priority or whether to give some classes greater significance than others. This helps determine the right metric to use.

2. Based on the number of instances per class, there are many factors to consider. It is necessary to check a dataset to see if it is imbalanced (there is more data in certain classes than others) or if it is balanced (there are about equal amounts of instances for each class).

1.5 CONCLUSION

This chapter has provided a detailed insight into pest detection and classification by using AI techniques. The ML and DL methodology includes the performance aspects that are compared from different perspectives. Even though there are many research results in the field of DL, there are still many problems like lack of real-time datasets, the problem of over fitting, and so on that need to be resolved. The model's complexity can be reduced in the future by optimizing the hyperparameters, and it might even be expanded to include other insect pest image analysis.

REFERENCES

1. Chen, Y.-S., Hsu, C.-S., Lo, C.-L.: An entire-and-partial feature transfer learning approach for detecting the frequency of pest occurrence. *IEEE Access* 8, 92490–92502 (2020)
2. Lee, W.-S., Alchanatis, V., Yang, C., Hirafuji, M., Moshou, D., Li, C.: Sensing technologies for precision specialty crop production. *Computers and Electronics in Agriculture* 74(1), 2–33 (2010)
3. TNAU: *Insect ecology & integrated pest management.* https://agrimoon. com/wp-content/uploads/Insect-Ecology-IntegratedPest-Management.pdf
4. Strange, R.N., Scott, P.R.: Plant disease: A threat to global food security. *Annual Review of Phytopathology* 43(1), 83–116 (2005)
5. Kasinathan, T., Singaraju, D., Uyyala, S.R.: Insect classification and detection in field crops using modern machine learning techniques. *Information Processing in Agriculture* 8(3), 446–457 (2021)
6. Dawei, W., Limiao, D., Jiangong, N., Jiyue, G., Hongfei, Z., Zhongzhi, H.: Recognition pest by image-based transfer learning. *Journal of the Science of Food and Agriculture* 99(10), 4524–4531 (2019)
7. Ahmad Loti, N.N., Mohd Noor, M.R., Chang, S.-W.: Integrated analysis of machine learning and deep learning in chili pest and disease identification. *Journal of the Science of Food and Agriculture* 101(9), 3582–3594 (2021)
8. Lee, S.H., Chan, C.S., Mayo, S.J., Remagnino, P.: How deep learning extracts and learns leaf features for plant classification. *Pattern Recognition* 71, 1–13 (2017)
9. Tsaftaris, S.A., Minervini, M., Scharr, H.: Machine learning for plant phenotyping needs image processing. *Trends in Plant Science* 21(12), 989–991 (2016)

10. Fuentes, A., Yoon, S., Park, D.S.: Deep learning-based techniques for plant diseases recognition in real-field scenarios. In: *International conference on advanced concepts for intelligent vision systems*, pp. 3–14 (2020). Springer

11. Kaggle. www.kaggle.com/datasets/simranvolunesia/pest-dataset

12. Durgabai, R., et al.: Pest management using machine learning algorithms: A review. *International Journal of Computer Science Engineering and Information Technology Research (IJCSEITR)* 8(1), 13–22 (2018)

13. Bakhshipour, A., Jafari, A.: Evaluation of support vector machine and artificial neural networks in weed detection using shape features. *Computers and Electronics in Agriculture* 145, 153–160 (2018)

14. Thenmozhi, K., Reddy, U.S.: Crop pest classification based on deep convolutional neural network and transfer learning. *Computers and Electronics in Agriculture* 164, 104906 (2019)

15. Khalid, S., Khalil, T., Nasreen, S.: A survey of feature selection and feature extraction techniques in machine learning. In: *2014 science and information conference*, pp. 372–378 (2014). IEEE

16. Iyapparaja, M., Sivakumar, P.: Detecting diabetic retinopathy exudates in digital image processing hybrid methodology. *Research Journal of Pharmacy and Technology* 12(1), 57–61 (2019)

17. Choras, R.S.: Image feature extraction techniques and their applications for CBIR and biometrics systems. *International Journal of Biology and Biomedical Engineering* 1(1), 6–16 (2007)

18. Gopalakrishnan, C., Iyapparaja, M.: Multilevel thresholding based follicle detection and classification of polycystic ovary syndrome from the ultrasound images using machine learning. *International Journal of System Assurance Engineering and Management* 1–8 (2021)

19. Kalel, D., Pisal, P., Bagawade, R.: Color shape and texture feature extraction for content based image retrieval system: A study. *International Journal of Advanced Research in Computer and Communication Engineering* 5(4) (2016)

20. Gopalakrishnan, C., Iyapparaja, M.: Detection of polycystic ovary syndrome from ultrasound images using sift descriptors. *Bonfring International Journal of Software Engineering and Soft Computing* 9(2), 26–30 (2019)

21. Kumar, S., Chauhan, E.A.: A survey on image feature selection techniques. *International Journal of Computing Science and Information Technology* 5(5), 644 (2014)

22. Iyapparaja, M., et al.: Effective feature selection using hybrid ga-eho for classifying big data siot. *International Journal of Web Portals (IJWP)* 12(1), 12–25 (2020)

23. Javapoint: *Support vector machine.* www.javatpoint.com/machinelearning-support-vector-machine-algorithm

24. Javapoint: *K nearest neighbour algorithm.* www.javatpoint.com/k-nearest-neighbor-algorithm-for-machinelearning

25. Javapoint: *Decision tree classification.* www.javatpoint.com/machine-learning-decision-treeclassification-algorithm

26. Javapoint: *Random forest.* www.javatpoint.com/machinelearning-random-forest-algorithm

27. Li, W., Chen, P., Wang, B., Xie, C.: Automatic localization and count of agricultural crop pests based on an improved deep learning pipeline. *Scientific Reports* 9(1), 1–11 (2019)

28. Xia, D., Chen, P., Wang, B., Zhang, J., Xie, C.: Insect detection and classification based on an improved convolutional neural network. *Sensors* 18(12), 4169 (2018)

29. LeCun, Y., Bengio, Y., Hinton, G.: Deep learning. *Nature* 521(7553), 436–444 (2015)

30. Krishnamoorthy, N., Prasad, L.N., Kumar, C.P., Subedi, B., Abraha, H.B., Sathishkumar, V.: Rice leaf diseases prediction using deep neural networks with transfer learning. *Environmental Research* 198, 111275 (2021)

31. LeCun, Y., Bottou, L., Bengio, Y., Haffner, P.: Gradient-based learning applied to document recognition. *Proceedings of the IEEE* 86(11), 2278–2324 (1998)

32. Veeragandham, S., Santhi, H.: A detailed review on challenges and imperatives of various cnn algorithms in weed detection. In: *2021 International conference on artificial intelligence and smart systems (ICAIS)*, pp. 1068–1073 (2021). IEEE

33. mygreatlearning:*Alexnet*.www.mygreatlearning.com/blog/alexnetthe-first-cnn-to-win-image-net/

34. Veeragandham, S., Santhi, H.: Effectiveness of convolutional layers in pretrained models for classifying common weeds in groundnut and corn crops. *Computers and Electrical Engineering* 103, 108315 (2022)

35. Olugboja, A., Wang, Z., Sun, Y.: Parallel convolutional neural networks for object detection. *Journal of Advances in Information Technology* 12(4) (2021)

36. mygreatlearning: *Vgg16.* www.mygreatlearning.com/blog/introductionto-vgg16/.

37. Nanni, L., Maguolo, G., Pancino, F.: Insect pest image detection and recognition based on bio-inspired methods. *Ecological Informatics* 57, 101089 (2020)

38. Kasinathan, T., Uyyala, S.R.: Machine learning ensemble with image processing for pest identification and classification in field crops. *Neural Computing and Applications* 33(13), 7491–7504 (2021)

39. Miranda, J.L., Gerardo, B.D., Tanguilig III, B.T.: Pest detection and extraction using image processing techniques. *International Journal of Computer and Communication Engineering* 3(3), 189 (2014)

40. Pattnaik, G., Parvathy, K.: Machine learning-based approaches for tomato pest classification. *TELKOMNIKA (Telecommunication Computing Electronics and Control)* 20(2), 321–328 (2022)

41. Mignoni, M.E., Honorato, A., Kunst, R., Righi, R., Massuquetti, A.: Soybean images dataset for caterpillar and diabrotica speciosa pest detection and classification. *Data in Brief* 40, 107756 (2022)

42. Fei, H., Fan, Z., Wang, C., Zhang, N., Wang, T., Chen, R., Bai, T.: Cotton classification method at the county scale based on multi-features and random forest feature selection algorithm and classifier. *Remote Sensing* 14(4), 829 (2022)

43. Dey, A., Bhoumik, D., Dey, K.N.: Automatic multi-class classification of beetle pest using statistical feature extraction and support vector machine. In: *Emerging technologies in data mining and information security,* pp. 533–544 (2019). Springer.

44. Cheng, X., Zhang, Y., Chen, Y., Wu, Y., Yue, Y.: Pest identification via deep residual learning in complex background. *Computers and Electronics in Agriculture* 141, 351–356 (2017)

45. Nurul Afiah Mohd Johari, S., Khairunniza-Bejo, S., Rashid Mohamed Shariff, A., Azuan Husin, N., Mazmira Mohd Basri, M., Kamarudin, N.: Identification of bagworm (metisa plana) instar stages using hyperspectral imaging and machine learning techniques. *Computers and Electronics in Agriculture* 194 (2022)

46. Turkoglu, M., Yaniko˜glu, B., Hanbay, D.: Plantdiseasenet: Convolutional neural network ensemble for plant disease and pest detection. *Signal, Image and Video Processing* 16(2), 301–309 (2022)

47. Chodey, M.D., Noorullah Shariff, C.: Hybrid deep learning model for infield pest detection on real-time field monitoring. *Journal of Plant Diseases and Protection* 1–16 (2022)

48. Nanni, L., Manf'e, A., Maguolo, G., Lumini, A., Brahnam, S.: High performing ensemble of convolutional neural networks for insect pest image detection. *Ecological Informatics* 67, 101515 (2022)

49. de Castro Pereira, R., Hirose, E., de Carvalho, O.L.F., da Costa, R.M., Borges, D.L.: Detection and classification of whiteflies and development stages on soybean leaves images using an improved deep learning strategy. *Computers and Electronics in Agriculture* 199, 107132 (2022)

50. Sabanci, K., Aslan, M.F., Ropelewska, E., Unlersen, M.F., Durdu, A.: A novel convolutional-recurrent hybrid network for sunn pest – damaged wheat grain detection. *Food Analytical Methods* 15(6), 1748–1760 (2022)

51. Suthakaran, A., Premaratne, S.: Detection of the affected area and classification of pests using convolutional neural networks from the leaf images. *International Journal of Computer Science Engineering (IJCSE)* 9(1) (2020)

52. Abirami, N., et al.: Protecting the farming land from insects damage to growing crops using deep convolutional neural network. *Turkish Journal of Computer and Mathematics Education (TURCOMAT)* 12(10), 1361–1366 (2021)

53. Rajalakshmi, D., Monishkumar, V., Balasainarayana, S., Prasad, M.S.R.: Deep learning based multi class wild pest identification and solving approach using cnn. *Annals of the Romanian Society for Cell Biology* 16439–16450 (2021)

54. Tetila, E.C., Machado, B.B., Astolfi, G., de Souza Belete, N.A., Amorim, W.P., Roel, A.R., Pistori, H.: Detection and classification of soybean pests using deep learning with uav images. *Computers and Electronics in Agriculture* 179, 105836 (2020)

55. Khanramaki, M., Asli-Ardeh, E.A., Kozegar, E.: Citrus pests classification using an ensemble of deep learning models. *Computers and Electronics in Agriculture* 186, 106192 (2021)

56. Chen, P., Li, W., Yao, S., Ma, C., Zhang, J., Wang, B., Zheng, C., Xie, C., Liang, D.: Recognition and counting of wheat mites in wheat fields by a three-step deep learning method. *Neurocomputing* 437, 21–30 (2021)

57. Alves, A.N., Souza, W.S., Borges, D.L.: Cotton pests classification in field-based images using deep residual networks. *Computers and Electronics in Agriculture* 174, 105488 (2020)

58. Wang, J., Li, Y., Feng, H., Ren, L., Du, X., Wu, J.: Common pests image recognition based on deep convolutional neural network. *Computers and Electronics in Agriculture* 179, 105834 (2020)

59. Li, Y., Wang, H., Dang, L.M., Sadeghi-Niaraki, A., Moon, H.: Crop pest recognition in natural scenes using convolutional neural networks. *Computers and Electronics in Agriculture* 169, 105174 (2020)

60. Cheng, X., Zhang, Y.-H., Wu, Y.-Z., Yue, Y.: Agricultural pests tracking and identification in video surveillance based on deep learning. In: *International conference on intelligent computing*, pp. 58–70 (2017). Springer

61. Wang, R., Zhang, J., Dong, W., Yu, J., Xie, C., Li, R., Chen, T., Chen, H.: A crop pests image classification algorithm based on deep convolutional neural network. *TELKOMNIKA (Telecommunication Computing Electronics and Control)* 15(3), 1239–1246 (2017)

62. Xin, M., Wang, Y.: Image recognition of crop diseases and insect pests based on deep learning. *Wireless Communications and Mobile Computing* 1–15 (2021)

63. Wang, X., Liu, J., Zhu, X.: Early real-time detection algorithm of tomato diseases and pests in the natural environment. *Plant Methods* 17(1), 1–17 (2021)

64. Tensorflow. www.tensorflow.org/

65. Keras. https://en.wikipedia.org/wiki/Keras

66. Caffe. https://caffe.berkeleyvision.org/

67. Pytorch. https://pyimagesearch.com/2021/07/05/what-is-pytorch/

CHAPTER 2

Integration of Artificial Intelligence (AI) and Other Cutting-Edge Technologies

S. Anitha and E. M. Roopa Devi

2.1 INTRODUCTION

The Internet of Things (IoT), a recent development that allows internet-based connections between electrical gadgets and sensors, was launched in order to develop lifestyles and cope with the fast-paced world. The term IoT refers to a paradigm in which every type of imaginable device is equipped with networking and computational capabilities. IoT intelligently connects objects.

IoT is a network of physical items, or "things," that are connected to sensors, software, devices, actuators, and other technologies to communicate and exchange data with other things with the help of the internet. IoT nodes are built into a variety of things, including mobile phones, industrial machinery, and wireless sensors. It has many applications, including smart cities [1], smart industries, smart transportation, smart buildings, smart food, and smart health sensing systems. It is used to strengthen the current technology and benefit farmers and society.

The model has lot of sensors in it to identify the temperature, humidity, speed, and pressure; to detect smoke; to gauge distance; and for other uses. The IoT gateway is used to send these inputs to a cloud server after the identification of inputs. Users can then access these data utilizing websites, mobile apps, and other means, as shown in Figure 2.1.

 DOI: 10.1201/9781003409502-2

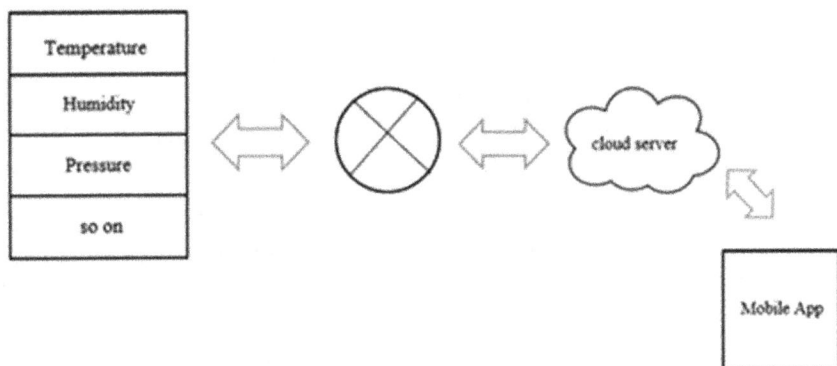

FIGURE 2.1 Model of different sensors transmitting information to a mobile app through the IoT gateway.

IoT applications improve daily lives with more recent Wi-Fi networks, advanced computational skills, and superior sensors, so the IoT will be the next frontier in the struggle for market dominance. IoT packages are predicted to add connectivity and intelligence to billions of everyday objects. It had been installed broadly. A few applications of IoT are as follows:

- Wearables
- Securities
- Smart cities
- Automation in agriculture
- Industrial automation
- Smart home applications
- Healthcare

IoT also allows for the improvement of the financial and industrial growth of a place, as well as trade change and stock change markets. A major challenge for IoT is the security of records and data. The internet is always going through protection threats and cyber assaults from hackers, and for this reason, the information and records are insecure. Hence, the main difficulty for IoT developers in exchange and economic systems is safety. So developers are working hard to develop a secure path.

The IoT facilitates humans' lifestyles and work. It permits machines to do the heavy lifting and tough duties and makes human life relaxed, wholesome, efficient, and healthy. As an example, an altogether linked device will change the entirety from normal: that is, if a snooze button is clicked, a unique sound makes the coffee machine turn on automatically; the refrigerator will find groceries on its own to cook; the oven will prepare meals according to the given menu for the day; the phone will schedule work for the day; and the automobile will robotically connect with the GPS, and at the same time as starting, it will check if gasoline is available or not and also analyze if everything is in secure mode or not. The developments in the IoT world are limitless.

A new kind of world is connected to the IoT. Almost all of the gadgets and equipment used are connected to a network. They can be used collectively to clear up complicated duties that require a high level of intelligence. The embedded sensors, actuators, CPUs, and transceivers found in IoT devices enable this intelligence and networking. IoT is not alone in technology; it is a group of various technologies that collaborate across platforms. Actuators and sensors are tools for interfacing with the physical world. To draw valuable conclusions from data that sensors have acquired, it must be stored and analyzed intelligently.

2.2 FUNCTIONAL BLOCKS OF AN IOT ECOSYSTEM

Since the IoT has a limitless variety of applications, countless coexisting IoT ecosystems are emerging. However, in the case of stripping the atmosphere down to its most fundamental components, a reliable schema will be discovered. An instrument collects data and transmits it over the network to a platform, which compiles the statistics for the agent to use later. Gadgets, networks, systems, and dealers are important factors in an IoT ecosystem. Sensing devices, processing devices, gateways, and apps make up the fundamental building blocks of an IoT device. Each of the nodes needs to possess specific characteristics in order to form a functional IoT machine.

2.2.1 Sensors

The primary goal of IoT devices is to gather information from their surroundings (sensors) or transmit data to others around them (actuators) so that they may be quickly identified across a wide network. These devices should be uniquely recognized and have entirely distinct IP addresses. These must be active, which entails that they ought to be able to gather real-time data. Depending on the user's demands, these can either function independently or through the user (consumer-managed). Temperature sensors, humidity sensors, oil pressure sensors, flow sensors, gas sensors, moisture sensors, and so on are a few examples of some sensors.

2.2.2 Processors

In an IoT system, the brain is its processor. Its primary function is to process the information obtained by sensors to separate useful information from the full volume of amassed raw data. In a single statement, it can be claimed that it gives data intelligence. Processors mostly work in real time and are used to manipulate applications. It is used for protecting the data that is being encrypted and decrypted. As microcontrollers, embedded hardware devices, and many others are connected to processors, these can be used to process the data.

2.2.3 Gateways

Gateways are used to route the processed data and send it to the right places where it's (data) correctly utilized. Alternatively, the gateway might enable data to be communicated back and forth. It displays data connectivity over a network. Any IoT system must have network connectivity in order to interact. Routers, WAN, LAN, and PAN are few notable examples of network gateways.

2.2.4 Applications

Applications are necessary for the proper use of all required data. Cloud-based programs are accountable for rendering powerful means to collect data. Packages are managed by users and serve as delivery points for specific services. Some examples of programs are home automation applications, protection systems, industrial automation hubs, and so on.

2.3 TECHNOLOGIES INVOLVED IN IOT DEVELOPMENT

IoT hardware devices can interact with various devices, packages, and cloud-based services thanks to networking technologies [2]. Modern protocols define the rules and codes that devices employ to build, manage, and transmit data over networks. Network construction involves the employment of a "stack" of technologies. At the bottom of the stack is Bluetooth LE generation. Range, bandwidth, power utilization, inconsistent connectivity, interoperability, and security are constraints to keep in mind when choosing which networking technologies to implement inside IoT applications.

Following are the few network technologies and brief summaries of each. A class of technology called as Low Power Wide Area Network (LPWAN) is designed for low-power, long-distance wireless communication. These are the ideas widely used for the installation of IoT devices, such as wireless sensor deployments.

LPWAN NB-IoT and LTE-M are the requirements for cellular that deals with issues, including shortage of electricity and low-cost IoT connectivity options in current mobile networks. A low-power version of the popular 2.4 GHz Bluetooth wireless communication is known as Bluetooth Low Energy (BLE). It is designed for communication across small distances (less than 100 meters), typically in the form of a star, with a single primary device controlling a large number of supporting devices.

ZigBee technology operates on a wireless communication spectrum of 2.4GHz. ZigBee has a long range of up to 100 meters, which is more than the range of BLE. ZigBee also has a slight decrease in recording rate of 250 kbps maximum in comparison to BLE at 270 kbps. For incredibly close communication (up to 4 cm), the near-field communication (NFC) protocol is utilized, along with maintaining an NFC tag or card near the reader. NFC is frequently used in payment systems; however, it can also be used for asset tracking through the use of smart labels during check-in procedures.

RFID is known as Radio Frequency Identification. Information and identifiers are stored on RFID tags. An RFID reader reads the tags after they have been affixed to objects. RFID characteristically has a range of less than a meter.

Wi-Fi is a kind of wireless networking with IEEE 802.11a/b/g/n standards. The maximum data speed is provided by 802.11ng, but this comes at the cost of excessive electricity consumption; hence 802.11b or g is the only option allowed for IoT devices to save power.

Ethernet carries out IEEE 802.3 standard and is broadly used for wired connectivity within LANs. Every IoT device does not require desk-bound Wi-Fi.

2.4 WORKING OF IOT

IoT operations are rather simple. First, it gathers knowledge about essential resources (names, addresses, etc.) and associated characteristics of items utilizing computerized identity and notion technologies such as RFID, wireless sensors, and satellite TV for PC positioning. In other words, the sensors, RFID tags, and all other uniquely identifiable items or "things" acquire real-time data (facts), with the distinctive feature of valuable smartphones featuring hubs. Second, beneath many varieties of communications technology, object-related information used in the information network provides services, intelligent indexing, and integration of the information associated with a large number of items. Eventually, utilizing sensible computing technology together with fuzzy, cloud computing and semantic analysis will analyze the data to make intelligent decisions [3].

2.5 IOT PLATFORM

The world of physical items and useful insights is connected via an IoT platform. By using a variety of tools and features, IoT platforms allow us to develop unique hardware and software products [4] for gathering, storing, managing, and analyzing the enormous amounts of data produced by connected assets and devices. The IoT platform aids in a better understanding of what the client needs, thus making it easier to create products. It gives businesses more operational intelligence and visibility, which facilitates better decision-making [5]. A multi-layer generation called an IoT platform enables simple provisioning, control, and automation of related devices in the IoT world. It generally links equipment to the cloud by means of connecting devices, agency-grade safety methods, and extensive data processing powers.

An IoT platform provides developers with a set of ready-to-use functionalities that vastly accelerate the creation of connected device applications while addressing scalability and pass-tool compatibility issues. Therefore, depending on how it looks, an IoT platform may be wearing unique hats. Middleware is a term used to describe the software that connects remote devices to user applications and manages all interactions between the hardware and application layers (or other devices). It is frequently called a "cloud enablement platform" or an "Internet of Things enablement platform," both of which highlight its key competitive advantage of enabling well-known devices to access cloud-based services and applications.

2.6 IOT PLATFORMS OVERVIEW

In IoT, concepts with Arduino and Raspberry Pi help in the emergence of a network of various devices communicating with one another and with their environment. Hardware solutions that assist developers in solving problems, such as creating autonomous interactive devices or conducting routine infrastructure-related tasks, are necessary for interoperability and connectivity.

2.6.1 Arduino

Arduino is smart for every environment with the aid of receiving source information from external sensors that is unique and is successful in communicating with other management factors spanning several appliances, motors, and drives [6]. Arduino has an integrated microcontroller that works using an Arduino software program. Tasks on this platform can be individual or collaborative, as determined by the use of add-ons and external equipment. The processing-based Arduino platform facilitates experimentation and attracts new users. The programming language is used to create lively interactive applications on the Java Virtual Machine (JVM) architecture.

There are many different types of Arduino boards, depending on the unique microcontrollers used. However, the Arduino IDE is one feature that unites all Arduino configurations. The range of inputs and outputs, velocity, working voltage, and shape element are the main factors influencing the discrepancies. Some boards are made to be embedded and don't have a hardware programming interface, while some require a minimum of 5V or less to operate.

2.6.2 Raspberry Pi

Raspberry Pi foundation and Broadcom have collaborated to develop Raspberry Pi of tiny single-board computers in the UK. The Raspberry Pi initiative started off with a focus on teaching computer science technology in classrooms and developing countries. Because of its affordability, versatility, and open structure, its miles are now broadly applied in the diffusion of fields and climate tracking.

2.7 WEARABLE IOT TECHNOLOGY

Although many IoT technology solutions depend on sensors and edge devices, wearable IoT devices cannot be disregarded. Wearable IoT gadgets like smartwatches, earphones, and extended reality (AR/VR) headsets are significant and will advance in the future.

Smart things affixed to clothing and other items, placed inside frames, and worn as accessories outdoors are known as IoT-enabled wearables. These devices have the ability to establish an internet connection in order to collect, send, and receive information to make wise decisions. Wearable computing, a key element of the IoT, is continually developing, moving from straightforward accessories to more complex, specialized goods. Smart wearables can communicate intelligibly and computationally with a wide range of devices, including smartphones. Smart wearable technology can boost productivity or security, improve quality of life, and optimize performance levels [7]. It has been witnessed that there is a rapid growth of smart wearable gadgets over the last few years, with each one tailored for different applications. Smartwatches, wristbands, headsets, earbuds, eye wears, hand-worn gadgets, foot and body straps, and clever jewelry are some of the wearable devices that have been advanced for specific programs, as depicted in Figure 2.2.

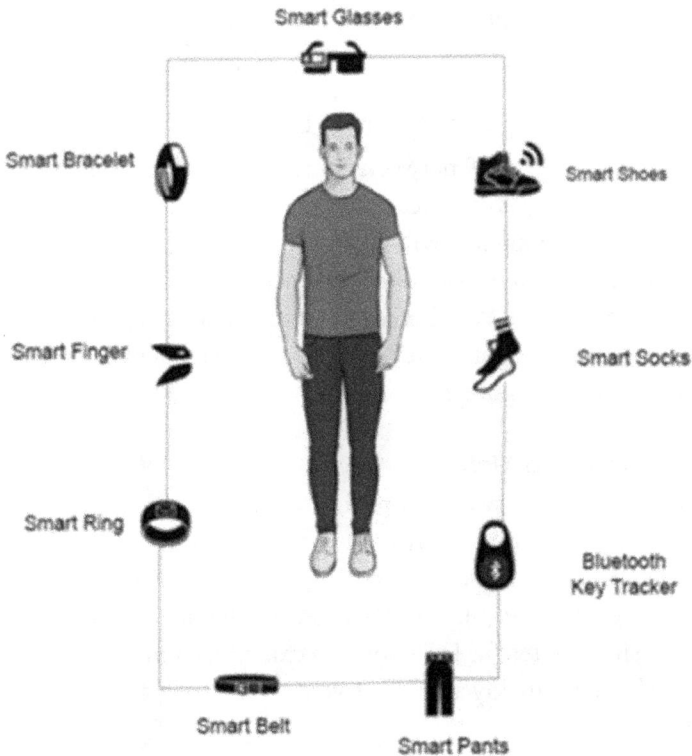

FIGURE 2.2 Different wearables developed for various applications.

Since wearable IoT devices run on batteries, electricity intake is considered a vital design element. Eventually, it does not imply that wearable devices don't perform complicated processing or because of the massive amount of energy used, the number of records that may be transferred is limited. Usually, if a wearable IoT gadget does a lot of processing or transfers a lot of information, its battery is required to be changed more regularly.

Once linked IoT systems and numerous issues with data proprietorship, laws on sharing records, privacy, and security are handled, the advantages of wearable IoT may be completely recognized. The majority of researchers explore IoT wearables as unlicensed relatively short-range communication technologies, in particular Wi-Fi and Bluetooth, to demonstrate the user's health, position, and safety. However, the usefulness of these responses is limited to connecting to the internet via a mobile device or gateway. Through the cellular IoT technology's development (e.g., LTE-M and NB-IoT) added in 3rd Generation Partnership Project (3GPP) Release 13, new strategies and programs are anticipated to be suggested for wearable IoT gadgets.

2.7.1 IoT Connectivity – 5G, Wi-Fi 6, LORA WAN

The major issue facing IoT networks in recent years has been the wireless data rate. Edge computing, wearables, smart homes, sensors, and other IoT technology components will advance with other technologies [8]. IoT solutions are more viable given that newer connectivity options have more infrastructures. LORA WAN, Wi-Fi 6, and other connectivity technologies fall under this category. The sequence of processes in IoT communication is illustrated in Figure 2.3.

2.7.1.1 5G: Advanced Mobile Networks

Many IoT technology systems are first deployed with a variety of edge devices, sensors, or other devices before it is maintained. However, in some situations, such as outdoor settings, mobile networks like LTE may be alternate possibility [9]. For 4G LTE, bandwidth is a limitation. The data processing needed for IoT networks can be accommodated by 5G networks much more quickly and effectively.

2.7.1.2 Wi-Fi 6

Wi-Fi in the 6 GHz band significantly boosts IoT technologies' bandwidth capability in indoor environments. If a network of devices communicates

FIGURE 2.3   Sequence of processes in IoT communication.

more quickly, it will be more dependable. Wi-Fi 6 may also be used in houses, making it a more beneficial technology for IoT in smart homes. Wi-Fi 6 enhances the rollout of 5G networks by assuring connectivity indoors, in public spaces and smart buildings, as well as in underserved areas, where 5G is much less successful. Wi-Fi 6 also offers improved performance, lower latency, and quicker data rates. Wi-Fi 6 has several features, including Orthogonal Frequency Division Multiple Access (OFDMA), to allow it to more effectively accommodate the different needs of IoT devices. With Wi-Fi 6 routers, any Wi-Fi 5 or older devices are fully backward compatible.

2.7.1.3 LoRaWAN

Low-Power, Wide Area Networking (LoRaWAN) protocol focuses on important IoT requirements like end-to-end security, bi-directional communication, localization services, and mobility for wirelessly attaching battery-powered "things" to the internet in local, regional, or global networks. LoRa technology has transformed IoT by providing power-efficient and long-distance data connectivity. Sensor devices with LoRa chipsets support a wide range of IoT applications by transferring packets with crucial data when linked to a non-cellular LoRaWAN network. Based on its extensive adoption, LoRa is the de facto IoT technology and will be used to connect billions of IoT devices. LoRa is adaptable for indoor or outdoor use cases in smart supply chains and logistics, smart utilities and metering, smart homes and buildings, smart environments, smart cities [10], smart agriculture, and industrial IoT (IIoT). Through the use of an increasing number of IoT vertical applications, LoRa devices and the LoRa WAN standard are enhancing lifestyles and boosting corporate efficiency.

2.7.1.3.1 Characteristics of LoRa-WAN technology

- Line of sight long-distance communication up to 10 miles.

- Using class A or class B mode on devices causes longer downlink latency. For longer battery life, extended battery can be used that has a life of up to 10 years.

- Low maintenance and device costs.

- Radio spectrum without licenses, but regional restrictions do apply.

- Low power consumption, although the payload size is constrained between 51 and 241 bytes, depending on the communication rate. The transmission rate varies between 3 Kbit/s and 27 Kbit/s, and the maximum payload size is 222 bytes.

2.7.2 AIoT – Artificial Intelligence and IoT Technology

Supporting AI software is the IoT technology's most common and fascinating use. Artificial intelligence and IoT complement each other. IoT gains from AI's distributed data as well as its improved administration [11].

2.7.2.1 Network of Data

Because artificial intelligence technologies are essentially data-driven, IoT sensors are significant in addition to the machine learning data pipeline [12]. By 2026, the market for AI in IoT technology would be worth $14,799 million, according to Research and Markets. High-quality data is essential for the effectiveness of machine learning techniques. Predictive maintenance uses real-time data from IoT sensors to predict when future equipment maintenance is required [13]. This is one of the most significant applications of AI in manufacturing.

2.7.2.2 Visual Inspection: AI and IoT Work Together

The benefit of IoT and AI working together applies to industrial and distribution sectors through visual inspection. Machine learning is good at finding patterns, but it needs high-quality data to do so. The opportunities for IoT networks and machine learning will become increasingly popular over the coming years [14].

2.7.2.3 Edge AI

Artificial intelligence depends heavily on data flow and the implementation of sophisticated machine learning algorithms. Edge computing creates a modern computing strategy that brings AI closer to the locations where data is generated and its computation occurs [15]. Edge AI may operate on a variety of hardware platforms, including basic MCUs and cutting-edge neural processing units, as shown in Figure 2.4. Machines and IoT devices are two types of edge AI hardware. Edge computing AI is a new field that has been created as a result of the convergence of AI and edge computing [16].

Characteristics of Edge AI

- **Reduced latency:** Edge AI lowers the delay by localizing data processing (at the device level), whereas data transfer from the cloud requires time.

- **Real-time analytics:** The main advantage of edge computing is real-time analytics. With the help of sensors and IoT devices, high-performance computing is brought by edge computing in those areas.

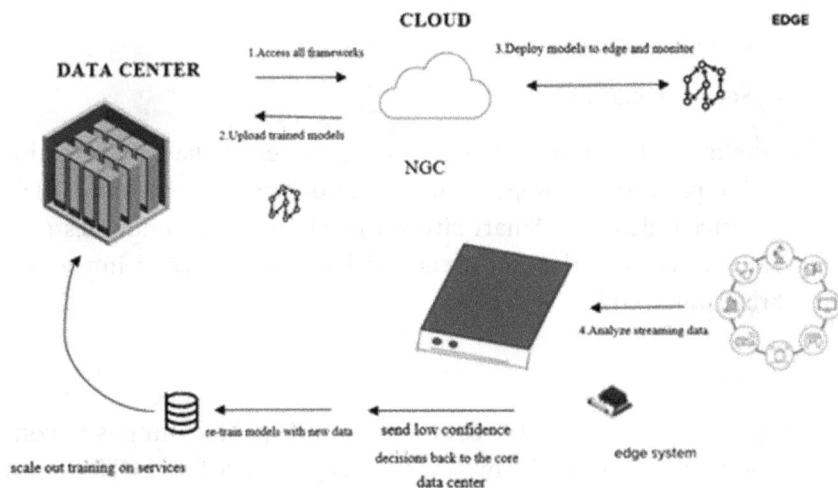

FIGURE 2.4   Edge computing with AI for IoT communication.

- **Higher processing speeds:** Processing time is much faster when the data is handled locally than by the cloud.

- **Reduced cost and bandwidth usage:** Since data is handled locally in the device itself in Edge AI, the internet traffic and cloud storage cost is reduced.

- **Enhanced data security:** The major part of processing data is done either locally or on the edge device itself. When the amount of data stored or transferred to the cloud reduces, it increases security in which data is hidden from internet predators.

- **Scalability:** Edge AI can handle finite quantity of data. There is no need to transfer data to the cloud server if video or image data from numerous sources needs to be processed simultaneously.

- **Increased dependability:** Edge AI System is more reliable due to increased speed and security standards.

2.8 APPLICATIONS

Introduction

Overall in the city planet, 50% of the population is living in cities and 70% of teh population is expected to be living in cities by 2050. Three major problems due to this transformation incur, which are:

- Environmental impact

- Economic growth

- Social evolution

Unemployment and protection to the resources in the cities are also major problems. Competitions are also increasing to show their industrial identities. Smart city with technology communities provides services to the end users and has environmental impact on urbanization [17].

Definition

The goal of the "smart city" concept in municipal planning is to control the assets of the city by securely integrating IoT and ICT. Due to rapid urbanization, cities occupy 2% of the earth's surface area and emit 73% of the world's greenhouse emissions. Around 600 urban centers account for 60% of the world's GDP. Smart cities could find solutions to the problems arising due to these infrastructures. Smart city increases employment and reduces environmental impact.

Working definition

- Manage and optimize existing and future infrastructure investments
- Provide more efficient services
- Support the city's and government's efforts to achieve their climate change adaptation and mitigation objectives.
- Make novel business models for providing services in the public and private sectors possible.

Examples:

- **Smart Transportation** – parking, traffic monitoring
- **Smart Healthcare** – remote monitoring, electronic record maintenance
- **Smart education** – Moodle, e-learning, allied campuses
- **Public safety and Security** – workflow for emergency assistance
- **Smart homes** – Meter reading, warning, lighting, and security devices
- **Roles, actors, engagements**

Cities have many actors. Actors have multiple roles. Each engages with infrastructure for various purposes.

TRANSPORT AND LOGISTICS

- Modes – Rail, Road, Air, Sea
- Used to deliver goods and services to end users
- Intra-urban – within city
- Inter-urban – city-to-city

Physical infrastructure

- **Manufacturer** – build infrastructure like bus, trains
- **Infrastructure managers** – buys infrastructure and implements, e.g., government
- **Operators** – frames schedules to operate and have customer relationship, e.g., Virgin trains in UK
- **Maintainers** – responsible for day-to-day running of infrastructure
- **End users** – who uses the infrastructure
- **Regulators** – provides rules and regulations, e.g., anti-monopoly act

Inputs include

- Traffic
- Working temperatures for infrastructure and operating conditions
- Subsidence
- Open data
- Corporate system
- OSS/BSS
- End Customers

Processing

- Multiple sources of data are combined.
- Used for decision-making like
  - How to provide service to low-income community
  - Information on public transportation during rush hours
  - Directions for rerouting traffic during significant incidents
  - Reducing serious train accidents

Packaging

- Creates information components like charts or traditional methods to communicate with end users from gathered inputs

- Large groups of end users may share the information packaging.

- Sometimes, sensitive information may be also shared, so they follow data sharing rules

- Packages must be easily understandable to make decisions

Distribution and Marketing

- Creation of information product

- Two types:

  - Improving internal decision-making

  - Product for resale to other economic actors

It is impossible to overestimate how the Internet of Things will affect contemporary culture. It has given us some means to communicate with surroundings and technology. Cities are attempting to become "smart" rather than simply being labeled as developed by employing this technology.

Smart towns use IoT gadgets like linked sensors, lights, and meters for data gathering and analysis. The towns make use of this data to improve, among other things, the public facilities, services, and infrastructure.

A smart city or IoT smart city has the following key attributes:

- Infrastructure for smart cities based on cutting-edge technologies

- Environmentally responsible actions

- A clever public transportation system

- Ability to enable citizen interaction with smart city ecosystems, connected buildings, mobile devices, etc. through integrated urban planning

IoT technology is the most important for smart cities. There are numerous IoT options for smart cities, ranging from internet-linked garbage cans and connected buildings to IoT-based fleet management [18]. IoT for smart cities enables local government authorities to remotely supervise and control linked equipment and guarantee efficient operations.

2.8.1 IoT for Smart Cities: Real-Time Examples

2.8.1.1 New York City

In an effort to reduce traffic-related fatalities, injuries from crashes, and infrastructure damage, NYC is testing an initiative involving linked vehicles [19]. The CTV infrastructure focuses mostly on safety applications, relying on communications between vehicles, infrastructure, and pedestrians.

2.8.1.2 Fujisawa

Only when sensors recognize a person, the street light gets turned on. Another significant issue is recycling. In Songdo, rainfall is collected, purified, and used to irrigate parks.

2.8.1.3 San Francisco

In order to improve safety, reduce collision rates, and speed up emergency vehicle response times, San Francisco has launched the Smart Traffic Signals Pilot, a test project that will look into the use of Multimodal Intelligent Traffic Signal Systems, Dedicated Short-Range Communication, Transit Signal Priority, and Emergency Vehicle Preemption technology.

2.8.2 IoT in Smart Home

A smart home automation application is a smartphone or tablet program that is used to virtually operate and administer connected non-computing equipment in the home. Several smart devices for various uses, including lighting, security, gardens, safety sensors, home entertainment, etc., make up IoT-based home automation systems. All of these gadgets are integrated over a single, gateway-created network and joined in a mesh network. Smart homes frequently use mobile phones to control internet-connected appliances and devices which is an important component of IoT.

2.8.2.1 Electricity

Now, the lighting in the house can be adjusted immediately based on the resident's requirements. To keep viewers from being distracted from the story, the lights might be set to automatically lower when a movie is about to begin. It is possible that when entering home, the lights will be turned on without needing to press a button.

If there is no one house, the system in house will automatically switch down the lights to save energy. All home's lights may be controlled by smartphone, laptop, and other connected devices.

2.8.2.2 Garden Areas

Sensors could be very useful for people who want to cultivate their own fruits and veggies at home. The right temperature and the amount of water and sunshine the plant is receiving can all be checked using the application by users. The software can keep track of the state of the earth, determine whether it has enough wetness, and, if required, activate a smart irrigation system. The sensor will automatically detect when the level of wetness is reached and switch off the irrigation system to conserve water.

2.8.2.3 Safety Sensors

Smart devices called safety sensors can spot issues in the home. They are able to take immediate action to avert problems and tell customers about them right away. They only require smartphone with internet access and sensors already installed in their home.

The air in home can be regularly monitored by temperature, humidity, and gas controllers, which can also send messages through the internet if the readings stray from the recommended range.

Fires, explosions, water breaches, and gas escapes can be prevented with the help of safety devices. A break-in attempt can be detected by proximity and video sensors, which then instantly trigger the alert and call the authorities.

2.8.3 IoT in Healthcare

Prior to IoT, patients and physicians were only able to interact orally, over the phone, and via text. It was impossible for medical professionals to constantly evaluate patients' health and give suggestions.

Thanks to IoT-enabled devices, remote monitoring in the healthcare sector is now feasible, unleashing the potential to keep patients secure and healthy and allowing doctors to deliver excellent therapy. IoT also has significant impact on improving treatment results and significantly reducing healthcare costs. Additionally, it raised the number of patients and improved interactions between the practitioner and patients [20].

With the aid of wearable devices like exercise bands and other electronically linked medical tools like blood pressure and heart rate monitor cuffs, glucometers, etc., patients can receive personalized treatment. The patient care process can be automated with the help of IoT by employing healthcare mobility technologies. The productivity of the healthcare industry has grown as a result of interoperability, connectivity between machines, and data movement [21]. With IoT connectivity, both patients and healthcare workers can save time. These devices can be set to prompt users to keep track of their blood pressure changes, meetings, appointments, and many other things.

Patient data gathering and records are revolutionized by IoT. The patient's electronic health record can safely keep the data gathered from their wearable gadget, which simplifies diagnosis because all the data is in one location.

2.8.3.1 Sensors
IoT in healthcare uses variety of sensors, including sphygmomanometers (blood pressure monitors), thermometers, EKG monitors, and pulse oximeters, to read patient conditions at any given time (records).

2.8.3.2 Connectivity
Better connectivity between devices or sensors from the microcontroller to the server and backend communication technologies (using Bluetooth, Wi-Fi, etc.) is made possible by IoT systems.

2.8.3.3 Analytics
In order to determine the patient's health parameters, the healthcare system analyzes data from sensors and correlates. Based on the results of the analysis, patient's health gets improved. IoT systems give healthcare practitioners access to data for all patients with full details on their monitor devices. This is depicted in Figure 2.5.

2.8.4 Automation in Agriculture
The primary goal of this project is to provide farmers with access to an automation system that addresses aforementioned problems like saving time, resources labor, and labor. The labor project serves as a comprehensive toolkit for the farmers, assisting them throughout the agricultural process. IoT is used in agriculture to monitor crops, survey and map fields,

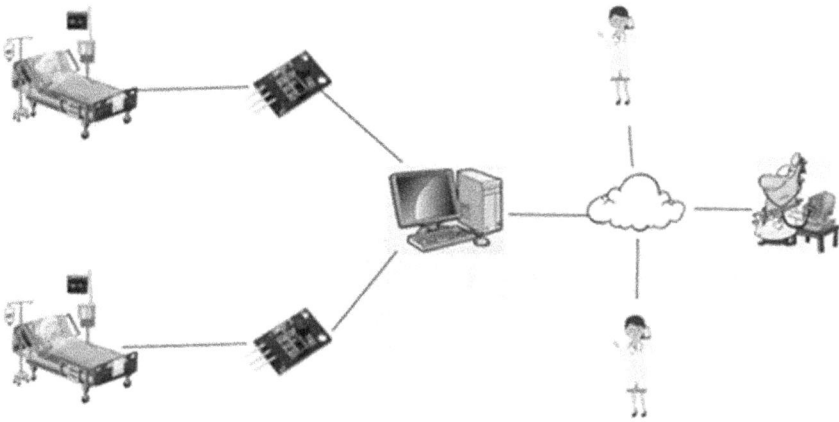

FIGURE 2.5   IoT in healthcare.

and provide farmers with the information they need to make time- and money-saving farm management decisions. These tools include robots, drones [22], remote sensors, computer imagery, and ever-evolving machine learning and analytical tools [23]. These sensor values are gathered and transmitted to the Raspberry Pi, which is running the Apache Web server, using an Arduino UNO or a custom-built MEGA. Also, Raspberry Pi is a storage or container that uses SQL database.

A communication channel has been created between the server and the sensor arrays using the ZigBee module. The farmer can access the server for getting information about the state of the field from anywhere at any time in order to save time and labor. To increase the communication's range, multi-hop communication is used. The sensor arrays' data are transferred to the neighbors through sensor arrays next to them through broadcasting. In this manner, after passing through numerous hops, the data finally arrives at the server. The major goals of this sort of serial transmission involving successive sensor arrays are to extend the total range and to ensure that all sensor array data is reliably received at the server end. There is a chance of data ambiguity and alternation, since sensor array in the multi-hop connection described above transmits both its own data and the data it receives from its neighbors.

The Figure 2.6 illustrates the sensor data are taken from the soil, concatenated, and communicated via a ZigBee module to either the server or through the sensor array, such as temperature-humidity sensor, sensor

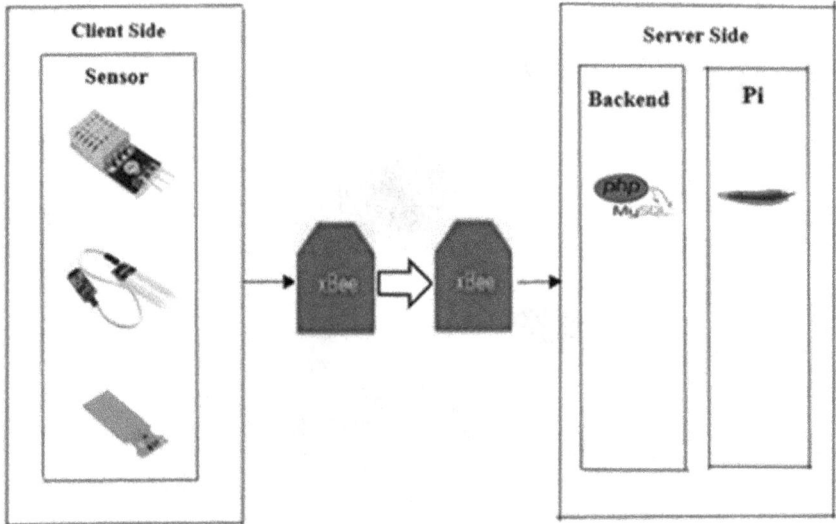

FIGURE 2.6   Client and server communication through xBee.

for soil moisture content, and depth detector for water levels. The subsequent sensor arrays receive these data and transmit them in order to enable multi-hop communication. Raspberry Pi in which the database is housed and the Apache Web server hosted receives the data.

2.8.5 Industrial Automation

The main objective of industrial automation is to reduce the demand for labor in the manufacturing process. This allows more efficient use of resources, industrial analytics, and quicker, safer output. In order to assign machines to work and automate certain process activities, the industrial process must be properly mapped out, and relationships between sub-processes must be understood.

IoT usage has significantly increased in industrial automation. Automation of various controlled devices, including computers and robotics, has aided in the management of numerous machines and processes. Thus, there is now less need for human involvement thanks to IoT. Performance has become more automated and under control as a result. IoT has aided businesses in boosting outcomes and operations, thus increasing productivity. The businesses have gained competitive edge thanks to the adoption of IoT technologies.

The phrase "Internet of Things" (IoT) describes a situation in which a growing number of IoT gadgets, such as smart houses, smart appliances, and industrial production machinery, are linked to the internet. These linked smart devices make it possible to manage processes across sectors.

2.8.6 IoT in Transportation

IoT in the transit industry uses a massive network of embedded sensors, actuators, smart items, and other cutting-edge gadgets. In order to provide helpful information, this network gathers data on the present situation and transmits it using specialized software. IoT in transit makes it safer, more environmentally friendly, and more practical than simply traveling from one place to another. For instance, a smart vehicle can handle several duties simultaneously, such as navigation, contact, amusement, and more dependable, effective transportation [24]. IoT enables travelers to maintain constant connectivity with all modes of transportation, as depicted in Figure 2.7. Several wireless technologies, such as Bluetooth, Wi-Fi, 3G, 4G, sophisticated traffic systems, and even other vehicles, are used to link the automobile to the internet.

By providing advantages like vehicle monitoring, safer route optimization [25], distance coverage, and others, IoT aids in the integration of the complete management processes. IoT is essential for the transportation

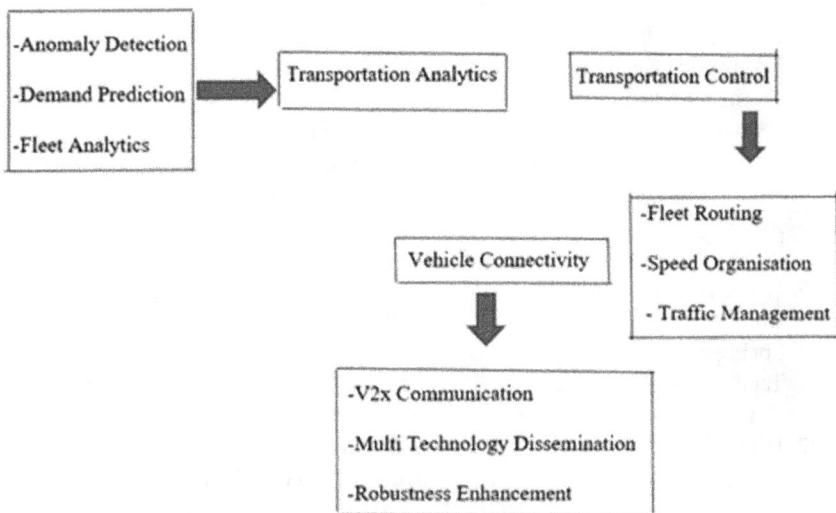

FIGURE 2.7 IoT in transportation.

sector and will be instrumental in managing operations to reduce adverse environmental effects. Growing awareness provides us with more flexibility and encourages the use of green technologies. The people can observe, assess, and enhance operations with the help of this new technology.

2.9 CONCLUSION

One of the most cutting-edge technologies currently in use is IoT. Everything is connected to the internet with IoT. IoT solutions are steadily growing, and IoT future trends will be crucial in the near future. Blockchain, Artificial Intelligence, 5G, Cloud Computing, and other significant technologies will be crucial to the development of the internet of things and global connectedness. Internet of Things (IoT) requires dependable connectivity, storage, and security because every linked object sends data packets. IoT presents a challenge to an organization's capacity to handle, oversee, and safeguard enormous quantities of data and connections from dispersed devices. As IoT develops at a rapid pace, efforts will be made to overcome obstacles, learn more about it, interact with it, and protect user privacy. Thus, IoT plays the most vital role in everybody's life.

REFERENCES

1. Alam, Tanweer, "Cloud-Based IoT Applications and Their Roles in Smart Cities", *Smart Cities*, vol. 4, no. 3, pp. 1196–1219, 2021, https://doi.org/10.3390/smartcities4030064
2. Wirtz, B. W., "Artificial Intelligence, Big Data, Cloud Computing, and Internet of Things", *in: Digital Government*, Springer Texts in Business and Economics, Springer, 2022, https://doi.org/10.1007/978-3-031-13086-1_6
3. Chakraborty, Sanjay and Lopamudra Dey, *Computing for Data Analysis: Theory and Practices*, Springer Science and Business Media LLC, 2023
4. Yazid, Yassine, Imad Ez-Zazi, Antonio Guerrero-González, Ahmed El Oualkadi and Mounir Arioua, "UAV-Enabled Mobile Edge-Computing for IoT Based on AI: A Comprehensive Review", *Drones*, vol. 5, no. 4, 2021, https://doi.org/10.3390/drones5040148
5. Cyber Intelligence for Business and Management Innovations, *International Conference on Applications and Techniques in Cyber Security and Intelligence*, Springer Science and Business Media LLC, 2018
6. Tsiatsis, Vlasios, Stamatis Karnouskos, Jan Höller, David Boyle and Catherine Mulligan, *Asset Management*, Elsevier BV, 2019
7. Dian, F. John, R. Vahidnia and A. Rahmati, "Wearables and the Internet of Things (IoT), Applications, Opportunities, and Challenges: A Survey", *IEEE Access*, vol. 8, pp. 69200–69211, 2020, https://doi.org/10.1109/ACCESS.2020.2986329

8. www.coursera.org/specializations/internet-of-things
9. Hua, Haochen, Yutong Li, Tonghe Wang, Nanqing Dong, Wei Li and Junwei Cao, "Edge Computing with Artificial Intelligence: A Machine Learning Perspective", *ACM Computer Survey*, vol. 55, no. 9, https://doi. org/10.1145/3555802, 2023
10. Belli, Laura, Antonio Cilfone, Luca Davoli, Gianluigi Ferrari, Paolo Adorni, Francesco Di Nocera, Alessandro Dall'Olio, Cristina Pellegrini, Marco Mordacci and Enzo Bertolotti, "IoT-Enabled Smart Sustainable Cities: Challenges and Approaches", *Smart Cities*, vol. 3, no. 3, pp. 1039–1071, 2020, https://doi.org/10.3390/smartcities3030052
11. Kaur, Gurjit, Pradeep Tomar and Marcus Tanque, *Artificial Intelligence to Solve Pervasive Internet of Things Issues*, Academic Press, pp. 103–123, 2021, ISBN 9780128185766, https://doi.org/10.1016/B978-0-12-818576-6.00006-X
12. Kamalam, G. K. and S. Anitha, "Cloud-IoT Secured Prediction System for Processing and Analysis of Healthcare Data Using Machine Learning Techniques", *Advanced Healthcare Systems*, 2022, https://doi.org/10.1002/9781119769293.ch10
13. https://xaltius.tech/artificial-intelligence-in-cloud-computing/
14. www.tantiv4.com/insights/iot-big-stories/the-difference-between-iot-ai-and-ml
15. https://pe.gatech.edu/courses/computing-for-data-analysis
16. Rath, Mamata, Jyotirmaya Satpathy and George Oreku, *Artificial Intelligence and Machine Learning Applications in Cloud Computing and Internet of Things*, IEEE, 2021, https://doi.org/10.1016/B978-0-12-818576-6.00006-X
17. Tsiatsis, Vlasios, Stamatis Karnouskos, Jan Höller, David Boyle and Catherine Mulligan, *Asset Management*, Elsevier BV, 2019
18. Petchrompo, S. and A. Parlikad, "A Review of Asset Management Literature on Multi-Asset Systems", *Reliability Engineering and System Safety*, vol. 181, pp. 181–201, 2019, https://doi.org/10.1016/j.ress.2018.09.009
19. Yazid, Yassine, Imad Ez-Zazi, Antonio Guerrero-González, Ahmed El Oualkadi and Mounir Arioua, "UAV-Enabled Mobile Edge-Computing for IoT Based on AI: A Comprehensive Review", *Drones*, vol. 5, no. 4, 2021, https://doi.org/10.3390/drones5040148
20. Siarry, Patrick, M. A. Jabbar, Rajanikanth Aluvalu, Ajith Abraham and Ana Madureira, *The Fusion of Internet of Things, Artificial Intelligence, and Cloud Computing in Health Care*, Springer, 2021
21. www.analyticsinsight.net/how-do-iot-and-machine-learning-go-hand-in-towards-a-smart-future/
22. Heidari, Arash, Nima Jafari Navimipour, Mehmet Unal and Guodao Zhang, "Machine Learning Applications in Internet-of-Drones: Systematic Review, Recent Deployments, and Open Issues", *ACM Computer Survey*, vol. 55, no. 12, 2023, https://doi.org/10.1145/3571728
23. www.bacancytechnology.com/blog/machine-learning-and-iot
24. Duarte, P. H. S., L. F. Faina, L. J. Camargos, L. B. D. Paula and R. Pasquini, "An Architecture for Monitoring and Improving Public Transportation

Systems", *IEEE 30th International Conference on Advanced Information Networking and Applications (AINA), Crans-Montana, Switzerland,* pp. 871–878, 2016, https://doi.org/10.1109/AINA.2016.39

25. Rajadevi, R., E. M. R. Devi, R. Shanthakumari, R. S. Latha, N. Anitha and R. Devipriya, "Feature Selection for Predicting Heart Disease Using Black Hole Optimization Algorithm and XGBoost Classifier", *International Conference on Computer Communication and Informatics (ICCCI)*, pp. 1–7, 2021, https://doi.org/10.1109/ICCCI50826.2021.9402511

A Detailed Case Study on Various Challenges in Vehicular Networks for Smart Traffic Control System Using Machine Learning Algorithms

Bandi Vamsi, Bhanu Prakash Doppala,

Mohan Mahanty, D. Veeraiah,

J. Nageswara Rao, and B.V. Subba Rao

3.1 INTRODUCTION

To address road accident issues, the Smart Traffic Control System (STCS) can provide early advice and efficient traffic planning. Internet-Based Vehicles (IBV) have been crucial to STCS [1]. By controlling traffic signals, gathering authentic network information, effectively communicating traffic information, reducing congestion, and directing transport participants, an IBV can aid in estimating the performance of a road infrastructure. We may predict the journey times of every passenger on the roadway in addition to assessing the traffic volume and awarding road guidelines on the IBV [2].

To help prevent traffic jams, STCS data can be gathered and analyzed to create a traffic prediction in advance [3]. The most effective technique

DOI: 10.1201/9781003409502-3

in the actual traffic connection management and security framework to identify communication networks and related traffic is through internet traffic classification [4].

Systems for network management and surveillance face difficulties when dealing with IoT data, which consists of heterogeneous wireless connections in different ways [5]. On the other hand, traffic data collection is a better approach that arranges traffic in accordance with different requirements and factors. Network security depends on the classification of network activity according to applications [6].

The categorization can be used for various data types, with an emphasis on network activity, including wireless services, IoT communication systems, smart buildings, and traffic conditions [7]. Data analysis, datasets, and traffic categorization attributes are crucial components of machine learning (ML) in a classification approach. The identification of network activity must take the collection, filtering, and ML approaches and their implementations into consideration [8]. By using the IBV network, the following list can be differentiated in order to estimate traffic flow:

- Time-to-time forecasting-based methods, like component influencing factors, indexing analysis, multiplier modeling, and additional designs

- Adaptive filtered techniques based on gradient descent, linear extrapolation, transformation adaptable area filtration, sub-band division, and so on can be used

- A number of strategies are dependent on dynamic systems, such as control algorithms and intelligent methods by using fuzzy and neural networks

Vehicle to Infrastructure (V-to-I), Infrastructure to Infrastructure (I-to-I), Vehicle to Person (V-to-P), and Vehicle to Vehicle (V-to-V) are the four main types of communication networks that are found in IBV systems, as depicted in Figure 3.1.

3.1.1 Benefits of Using STCS

The benefits of the STCS network according to the four kinds of communication networks are as follows:

1. From V-to-P network, we can establish a logical connection between different types of vehicles to person or own vehicle to person. If any

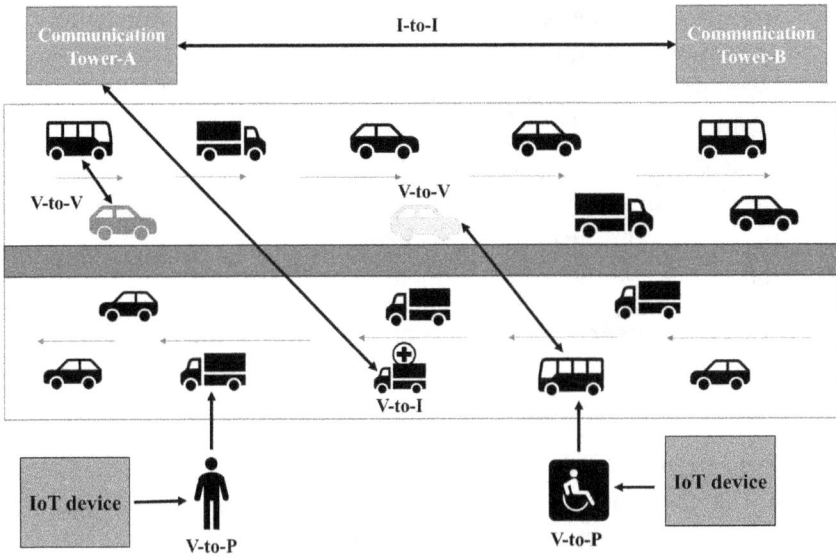

FIGURE 3.1 Internet-based vehicles communication system.

disabled person wants to get into a public transport vehicle, immediately he/she can send a request alert to the vehicle according to the
distance. If the vehicle has seat facility or availability, then the driver
accepts the request. Based on the location of the person, the vehicle
automatically slows down.

2. If any vehicle on the road crosses the speed limit, then a notification
alert will be sent to the police patrol vehicle. This can be done by
using the V-to-V category. Based on this alert, the police can send a
warning alert/message to the vehicle driver/vehicle owner directly to
reduce their speed.

3. Assume that there was a vehicle breakdown on the road. In this situation the person can send a request alert to public transport with
their current location. If any public transport vehicle is available in
that area, then the person can avail the transport, or else an emergency alert can be sent to the nearest police station. This can also be
done by V-to-V communication.

4. On the road, all types of vehicles pass by, including public, own,
and ambulances. Assume an ambulance is on the road, which indicates there is an emergency situation. To handle these types of situations, we need to give complete free space on the road to give way

to ambulances. Once the free space is ready, then the ambulance can go without any traffic and keep all other vehicles to one side. These types of emergency situations can be handled by a V-to-I communication system. By using this type of communication, the ambulance driver will send an emergency alert to available towers. Now the communication towers send an alert to all available vehicles on the road according to the region's coverage. Hence, we can control all the vehicles on the road without any congestion.

5. All other types of communication that fall into categories are regulated and communicated with one another among people and vehicles employing an I-to-I communication system.

6. The categorization of incoming traffic entices our curiosity and offers a wide range of vital uses for administration. Here, categorization is utilized to keep track of a variety of behaviors.

Traffic monitoring can be utilized to monitor and control safety while also evaluating the performance of the network. As a result, network traffic analysis using particular ML techniques is crucial for improving network operations and security. Improved methods are needed to recognize breaches, assess malware action, classify networking traffic information, and identify the different additional security features in order to handle increased artificial intelligence (AI) incoming traffic. ML shows the capacity to successfully resolve network issues.

3.1.2 Reinforcement Learning (RL)

Technologies that use ML depend on data and knowledge. These systems have a benefit compared to others in considerations of pre-processing, strategic planning, and learning period since they adapt to the surroundings and produce appropriate findings based on current variables without the need for additional quantitative or empirical investigations. Once an ML model has learned the data, it can reasonably forecast outcomes for comparable circumstances [9]. Reinforcement learning (RL) based techniques in ML could be the most effective for resolving this issue. Furthermore, traffic light control management challenges involve a variety of characteristics and an ability to adapt quickly [10].

Due to how they operate, RL techniques can offer the best solutions in this area. Smart traffic signal management issues have a significant amount of potential for RL because it can train and get better with practice. The RL technique makes sure that the surroundings are constantly managed and that the ML model improves by picking the best course of action. Since it acts in the environment for a system based on an exploratory process and produces some incentive, it gains from its experiences [11].

It develops a model that gains experience and knowledge and improves in accordance with the incentive or penalty process over time. A compensation is a crucial indicator of a model's effectiveness. The number of vehicles that wait for a signal or junction, the volume of an intersection or merged junction, standing duration, or a mixture of these variables are all possible possibilities for reward criteria. For instance, average wait time and other time characteristics perform better if they are fewer. However, a greater flow or quantity of moving autos is preferable. As a result, real-time congestion is better suited for RL model solutions since it can handle surprises more effectively. Additionally, it doesn't call for prior workplace familiarity or outside oversight [12].

Whenever we think of the flow of traffic, people don't just mean managing traffic load in a certain situation. Emergency situations or other types of unplanned interruptions might affect how traffic moves. Since they frequently require unique actions and state descriptions, these problems are considered outlier components of the road network in this perspective. Emergency rescue action time is a crucial concern, both in dispatching time and arriving at the event location, similar to the demanded flow of traffic. Selecting a reliable response system for emergencies is crucial since emergency services are given preference in traffic conditions [13].

3.1.3 Traffic Light Management System (TLMS)

A Traffic Light Management System (TLMS) should be able to recognize the level of priority and alter the ordering or conditions of the traffic signals to address the emergency situation. Other possible enhancements and factors to think about include route design, signaling junction light control, and positioning vehicles in areas covered in much more effective positions. Obviously, analyses include setup and solution criteria such as traffic load, speed limit, wait period, delay, capacity, and the number of stops. The major purpose of the traffic signal setup process is to give a junction improved phase timings and phases order. Understanding the timing protocol along with the lighting hierarchy must therefore be a priority [14].

Experience-based learning is possible with reinforcement learning. The reward description and state-action link are sufficient for this strategy, which does not need a predetermined model or collection of possibilities in order to respond to experiences. RL is a useful option to take into consideration because congestion is a dynamic situation and it can be difficult to design a proper management policy. In this context, researchers have concentrated more on RL-based light control issues.

As was previously noted, traffic light control places a strong emphasis on variables including light duration, phase order, and duration. Various research studies take into account the time of the traffic lights and the length of the green light. The primary goal of this method is to determine the ideal traffic signal frequency for a signal. Each light in the cycle could have the same number or a different number. Another significant factor that draws studies is phase sequence [15].

3.1.4 Machine Learning (ML)

ML methods concentrate on vehicular networks for congestion classification tasks. So, utilizing the actual IBV's flow, the current study examined the effects of each instance of ML dataset characteristics on classification results. Various network tasks, such as security and surveillance, defect identification, transport planning, and quality of service (QoS) characteristics, require performance characteristics of vehicular networks. It is crucial to categorize the network activity of IBVs because of their explosive growth. It can be difficult to solve the problem of IBV network traffic creation, particularly when malicious and benign data are needed to fit the predictions. The effectiveness of several ML algorithms linked to Intrusion Detection System (IDS) on IBV-based vehicular communications is compared in this research [16].

3.1.5 Traffic Monitoring Using Drone Technology

The exceptional capability of drone technology to remarkably respond to a broad range of unexpected conditions has recently led to its emergence. Autonomous aerial vehicles could also be used successfully in a number of situations, such as relief and search operations, surveillance equipment, agricultural production, and so on. The technical and practical advantages include low progression, the capacity to enlarge wireless connectivity zones, as well as the capacity to reach locations unachievable by humans. Numerous network configurations, including congestion, bandwidth, connectivity, and durability, are made more effective by drones. However, the use of drone systems raises a number of problems, such as the electromagnetic medium inherent unreliability, good mobility levels, and battery capacity, each of which can cause rapid variations [17].

This drone technology can therefore also be used to facilitate the design of road improvements, the creation of more effective public transportation networks, the determination of the best places for bicycle lanes, and many other eco-friendly transport methods. By combining these activities, the drones may be able to track current traffic while avoiding potential congestion.

3.2 LITERATURE REVIEW

Wahab et al. [18] state that the goal of network congestion control is to categorize traffic analysis in a way that indicates how to operate an intelligent traffic system and evaluate network security and efficiency procedures. This work describes various ML techniques for traffic monitoring and AI techniques to find and examine malware behavior. The classification and identification of the transportation system are becoming more and more dependent on high-speed transmission capacity. The method adopts a novel approach to modify the Naive Bayes algorithm's conditional independence assumption states that features are independent of each other given the class. It integrates a number of elements, including computation, a data, a connection, and navigation systems, to build a system for the flexible design of data centers. For automotive ad hoc networks, techniques like feature extraction are first used to remove unnecessary anomaly information and cluster intelligent data. For prediction accuracy and efficiency, a computerized mutual information filter outperforms the intermediate filter method.

Rani et al. [19] noted that for forecasting and detecting congestion problems in the smart vehicular networks, novel methods have been created for IBVs. An ML-based system using decision tree (DT), extra tree (ET), random forest (RF), and XGBoost methods is proposed for the IBV for transportation network activity for the smart traffic environment. According to ensemble methods and feature combining, this work's findings demonstrate that the ML model offers excellent detection performance and reasonable computing costs. For IOV-based vehicle network activity, feature-based and tree-based ML algorithms outperformed those without selecting features. The stacked method has an accuracy of classification of 99.05%, which is greater than the SVM performance of 98.01% and the KNN accuracy of 96.6%.

Saleem et al. [20] note that traffic jams and the frequency of accidents on the highways have increased due to the quick increase in the number of vehicles on the highway. Vehicle networks (VNs) have created numerous innovative concepts, such as communication between vehicles, routing, and traffic monitoring, to address this problem. In order to reduce traffic jams in urban cities, this study presents a fusion-based adaptive road congestion management system for virtual networks that uses ML approaches to gather traffic information and schedule traffic on accessible routes. The planned system offers vehicle operators cutting-edge technologies that let them watch the flow of traffic and the number of vehicles on the road directly, with the goal of preventing traffic bottlenecks. This model lessens congestion and enhances road traffic with an efficiency of 95% and a failure rate of 5%.

Kiani et al. [21] note that the solutions in the studies conducted just recommend that the state move on to the following phase when an urgency vehicle moves. Based on actual circumstances, this work suggests an innovative method for dealing with emergencies at intersections. Instead of just repeating to the following stage, the proposed approach identifies superior options for various situations using a combination of situation and RL approaches to identify the appropriate course of action for resolving emergency situations. In order to select a workable course of action from a collection of options, this algorithm employs a Q-learning-based system that trains from the traffic environment for an urgent situation. The recovery measures are divided into three categories: maximum, minimum, and average. According to testing results, the suggested strategy beats the typical single option recovery action-based strategy by about 80%.

Qiu, D. et al. [22] note that modern electrical networks have been found to incorporate more energy from renewable sources, which has significant advantages for climate change mitigation and expediting the low-carbon transformation. The inconsistent and unpredictable character of renewable energy, therefore, poses additional difficulties for the design and management of power sources. The Vehicle-to-Grid (V2G) technique has increasingly come to be regarded as a viable response to these problems in order to provide different service supplies for electricity systems. Because of the complexity of energy systems, such type of approach cannot manage the dynamical and unpredictable situation adequately. Recently, there has been an increase in interest in using cutting-edge RL techniques to address various electric vehicles (EV) scheduling problems, which has produced numerous good research publications and significant conclusions. This work highlights how these sophisticated methods can be used to solve a variety of EV dispatching challenges and includes a thorough assessment of well-known RL algorithms divided into single and multi-agents.

Wang et al. [23] note that routing protocol is a concept in network topologies and complies with the strategic transformation concept. It does not only satisfy requirements but also improves network security and efficiency. The size of the forwarding knowledge base and also the price of network elements are both significantly increased by the implementation of routing algorithms. This study includes two-dimensional networking to achieve the light implementation of multi-path navigation with a distributed network. Two-dimensional packet switching combined with SRv6, together with the accompanying SRv6 headers reduction technique, is a shared storage area. This study introduces segmented sequencing on the data layer, which can distribute at various ingress routers and is influenced by a way to capture large-area networks. After the input gateways release the routing path into the hierarchy, the intermediary routes the packages forward to categorize different type transmissions connectivity potential applications and to explore various reduction strategies. According to the experimental data, routing can cut down two-dimensional entries by 69%, as well as the overall average reduction ratio can increase to 70%.

Ayyub et al. [24] note that vehicular networks, a type of ad hoc network that employs automobiles as base stations and a vehicle-to-anything connectivity mechanism, are about to be deployed. One of the more crucial methods for improving durability, dependability, and extensibility of networking is clustering. Moreover, clustering uses frequency minimization to reduce the issues of hidden units by cutting down on excess and transmit delay. Clustering in vehicular communications still lacks a comprehensive strategy. In this context, we provide a thorough analysis of the most recent developments in vehicular ad hoc clustering techniques to research the existent clustering methods while considering clustering, development, and monitoring. We also give a thorough set of criteria for choosing centroids as well as the function of cognitive computing in the maintenance of cluster.

Park S. et al. [25] note that vehicle networks have gained a great deal of attention in the automotive sector and fifth-generation (5G) networking technology's convergence has accelerated due to its significant potential to increase both safety and efficiency. By using the deep network and SVM classifiers, the data representation is first implemented for each vehicle environment and then collaboration connections between several vehicular communications are carried out to provide an optimum network. A collaborative deep SVM classification method integrating multi-agent systems and collaborative networking has been developed for the improvement of this poor detection reliability issue in conventional vehicular intrusion detection systems. These outcomes demonstrate that, when applied to several vehicular communications, the data description approach can optimize the false negatives to 52% with an overall detection performance of 91%, outperforming the conventional single vehicular communication detection approach.

Wang et al. [26] note that the features of the dynamic situation and the quality of services (QoS) in a network domain must be taken into account for IoT service situations. Three concerns need to be handled by a roadway unit serving as a fog node in order to maximize delivering service rates while satisfying QoS parameters. The arrangement of operations with various effective times is the primary problem. The second approach takes into account restricted data storage and replaces the roadway unit cache. The next is communication collision prevention for networks shared by numerous vehicles that is dependent on the quality of service. By using this work, a type of Deep QoS Network for deep reinforcement training approaches can resolve these three problems.

A comprehensive collaborative strategy must be developed for both V-to-I and V-to-V linkages. The achievement of the V-to-I connections delays criteria while reducing the latency in edge knowledge gathering for V-to-V links. To be more precise, two problems, a mean delay reduction issue and maximum delay reduction individual problems, are developed to boost the overall network efficiency and guarantee that each user is treated fairly. These two issues are resolved using a multi-agent RL framework, and a new reward mechanism is suggested to assess the benefits of the different optimization goals in a single framework. Then, utilizing the common global communications reward, a regional strategy optimization method allows for each V-to-V user to acquire their own policy. A summary of the remaining survey works is listed in Table 3.1.

3.3 MATERIALS AND METHODS

In this section, various machine learning models like decision tree, random forest, and bagging techniques are discussed. The statical analysis for the datasets used in this work was analyzed with the respective parameters.

3.3.1 Dataset

In this work, we have used datasets for road traffic analysis named "local authority traffic" [33] and "region traffic" [33], which contain various parameters such as "kilometers," "miles," "car details," "all motor

TABLE 3.1   Summary of Remaining Related Works [27–32]

Author	Approach	Methodology	Summary
Ahmet et al., 2022 [27]	Machine Learning	1. Centralized learning (CL) 2. Federated learning (FL)	1. The adoption of FL versus CL in application for automotive networks to create ITS. 2. A thorough investigation into the viability of FL for machine learning-based vehicle networks, in addition to a look into object recognition using image datasets. 3. The main issues are data labelling and model construction from a knowledge – based perspective as well as available bandwidth, dependability, transport latency, security, and resource planning from a networking perspective.
Hyunhee Park et al., 2022 [28]	Deep Learning	Vehicle-to-Everything (V-to-X)	1. This work discussed about a traffic safety application-specific lightweight authentication framework built on the edges and utilizing deep learning. 2. Vehicles that are actually apart can create an automotive cloud where V-to-V interactions can be safeguarded due to authentication framework. 3. Depending on the total number of cars participating in IDS employing control area network activity, the result of the F1-score for this model range from 94.51 to 99.8%.
Samir et al., 2021 [29]	Internet of Vehicles (IoV)	1. Intelligent Traffic Management System 2. Smart Traffic Signal 3. Vehicular Ad-hoc Networks (VANET)	1. Vehicles can wirelessly connect with people nearby as well as a specific architecture because of VANET and IoV. 2. A modern traffic control network built on top of the VANET and IoV which is appropriate for Smart City applications and upcoming transportation systems.

(Continued)

TABLE 3.1 (Continued)

Author	Approach	Methodology	Summary
Mamoona et al., 2022 [30]	Internet of Things (IoT)	1. Smart traffic management system 2. Cloud computing	1. Huge volume of data gathered by different traffic sensors needs to be properly stored and used for effective traffic monitoring in order to control road congestion. 2. By deploying roadside communication units to provide actual traffic data on congestion and unscheduled traffic occurrences, this strategy promotes mobility.
Adil Hilmani et al., 2020 [31]	Internet of Things (IoT)	Wireless Sensor Network	1. A WSN-based intelligent transport system that gathers information on parking availability along with traffic flow in a smart city. 2. With the use of an Android application, this model cutting-edge features let drivers observe the amount of congestion and parking places nearby from a distance. 3. To simplify the process for travelers to search for a free parking spot so they may avoid making unnecessary journeys, to escape traffic congestion, to take a different route to prevent being trapped, and more.
Ahmad Naseem et al., 2022 [32]	Fog Computing	Vehicle-to-Everything (V-to-X)	1. Fog processors are positioned close to the roadside equipment in order to quickly process jobs that have been unloaded. 2. A knapsack scheduling technique is suggested to handle the offloaded workloads in the best way possible.

TABLE 3.2   Statistical Analysis of Local Authority Traffic

| | One-Sample Statistics | | | | |
Attribute Name	N	Mean	Std. Deviation	Std. Error Mean	P-value (2-tailed)
local_authority_id	5529	103.95	59.894	0.805	<0.000
Year	5529	2006.03	7.792	0.105	<0.000
link_length_km	5529	1914.89	2250.13	30.26	<0.000
link_length_miles	5529	1189.85	1398.16	18.80336	<0.000
cars_and_taxis	5529	118659683	132394926	17805248	<0.000
all_motor_vehicles	5529	148939763	166792422	22431226	<0.000

TABLE 3.3   Statistical Analysis of Region Traffic

| | One-Sample Statistics | | | | |
Attribute Name	N	Mean	Std. Deviation	Std. Error Mean	P-value (2-tailed)
Year	1645	2006.22	7.880	0.194	<0.000
region_id	1645	6.16	3.185	0.079	<0.000
road_category_id	1645	3.62	1.725	0.043	<0.000
total_link_length_km	1645	6436.12	11138.14	274.61	<0.000
total_link_length_miles	1645	3999.2	6920.92	170.64	<0.000
pedal_cycles	1645	46760807	84150860	2074796.9	<0.000
two_wheeled_motor_vehicles	1645	47573386	49942118	1231357	<0.000
cars_and_taxis	1645	398826375	301230327	74270396	<0.000
buses_and_coaches	1645	49491777	47215209	1164123.2	<0.000
Lgvs	1645	643201118	492137640	12133989	<0.000
all_hgvs	1645	277475760	292966667	7223293.4	<0.000
all_motor_vehicles	1645	500600580	371371257	91564122	<0.000

vehicles," "bus details," etc. [33]. The summary of the dataset "local authority traffic" is given in Table 3.2, and the dataset "region traffic" is given in
Table 3.3. The preview of these datasets from the IBM SPPS tool is shown
in Figures 3.2 and 3.3.

Histograms are similar representations of bar charts. Each bar represents the distribution of data in a particular group, rather than comparing
the groups. Also, every bar denotes a sequence range of data, which is the
equal number of frequencies for a particular point in the data. The histogram representation of all attributes of the "local authority traffic" dataset
is shown in Figure 3.4. The histogram representation of all attributes of the
"region traffic" dataset is shown in Figures 3.5 and 3.6.

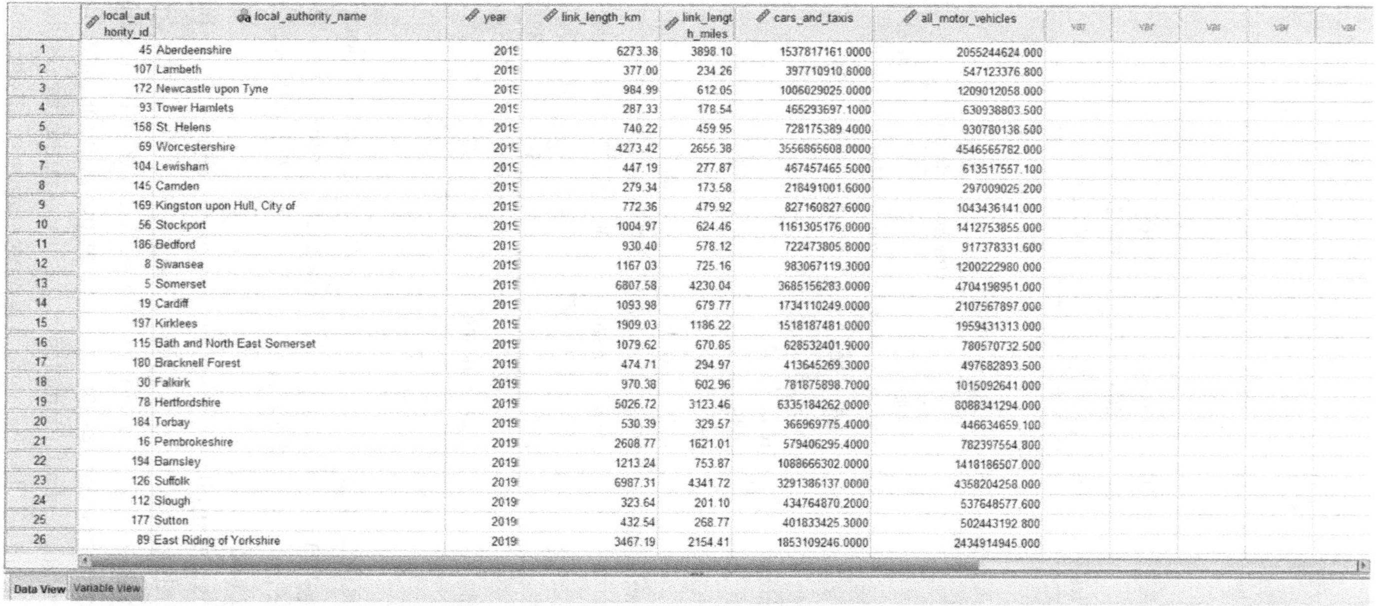

	local_authority_id	local_authority_name	year	link_length_km	link_length_miles	cars_and_taxis	all_motor_vehicles	var	var	var	var	var
1	45	Aberdeenshire	2019	6273.38	3898.10	1537817161.0000	2055244624.000					
2	107	Lambeth	2019	377.00	234.26	397710910.8000	547123376.800					
3	172	Newcastle upon Tyne	2019	984.99	612.05	1006029025.0000	1209012058.000					
4	93	Tower Hamlets	2019	287.33	178.54	465293697.1000	630938803.500					
5	158	St. Helens	2019	740.22	459.95	728175389.4000	930780138.500					
6	69	Worcestershire	2019	4273.42	2655.38	3556865608.0000	4546565782.000					
7	104	Lewisham	2019	447.19	277.87	467457465.5000	613517557.100					
8	145	Camden	2019	279.34	173.58	218491001.6000	297009025.200					
9	169	Kingston upon Hull, City of	2019	772.36	479.92	827160827.6000	1043436141.000					
10	56	Stockport	2019	1004.97	624.46	1161305176.0000	1412753855.000					
11	186	Bedford	2019	930.40	578.12	722473805.8000	917378331.600					
12	8	Swansea	2019	1167.03	725.16	983067119.3000	1200222980.000					
13	5	Somerset	2019	6807.58	4230.04	3685156283.0000	4704198951.000					
14	19	Cardiff	2019	1093.98	679.77	1734110249.0000	2107567897.000					
15	197	Kirklees	2019	1909.03	1186.22	1518187481.0000	1959431313.000					
16	115	Bath and North East Somerset	2019	1079.62	670.85	628532401.9000	780570732.500					
17	180	Bracknell Forest	2019	474.71	294.97	413645269.3000	497682893.500					
18	30	Falkirk	2019	970.38	602.96	781875898.7000	1015092641.000					
19	78	Hertfordshire	2019	5026.72	3123.46	6335184262.0000	8088341294.000					
20	184	Torbay	2019	530.39	329.57	366969775.4000	446634659.100					
21	16	Pembrokeshire	2019	2608.77	1621.01	579406295.4000	782397554.800					
22	194	Barnsley	2019	1213.24	753.87	1088666302.0000	1418186507.000					
23	126	Suffolk	2019	6987.31	4341.72	3291386137.0000	4358204258.000					
24	112	Slough	2019	323.64	201.10	434764870.2000	537648577.600					
25	177	Sutton	2019	432.54	268.77	401833425.3000	502443192.800					
26	89	East Riding of Yorkshire	2019	3467.19	2154.41	1853109246.0000	2434914945.000					

Data View
Variable View

FIGURE 3.2 Preview of local authority traffic dataset from the SPSS tool.

year	region_id	region_name	road_category	road_category_description	total_link_length_km	total_link_length_miles	pedal_cycles	two_wheeled_motor_vehicles	cars_and_taxis	buses_and_coaches	lgvs
2019	1 South West	4 PA	Class A Principal road	4253.7000000	2643.13	54356578.3200000	106150950.40000	8525295298.0	67183895.590	1601046776.000
2019	5 North West	4 PA	Class A Principal road	3793.0000000	2356.86	62731632.0700000	72243953.52000	9158518522.0	90102061.690	1452719635.000
2019	3 Scotland	3 TA	Class A Trunk road	2952.5000000	1834.60	4618188.70200000	50551466.39000	5002906792.0	54978483.220	1090807800.000
2019	10 West Midlands	2 PM	M or Class A Principal Motorway	3.8000000	2.36	.00000000	244706.17470	66100846.130	554004.47640	9836291.603
2019	6 London	3 TA	Class A Trunk road	1.6000000	.99	17228.75221000	108747.22740	9981850.014	5335.73582	1933932.336
2019	6 London	1 TM	M or Class A Trunk Motorway	60.3000000	37.47	.00000000	9650569.56700	1151238267.0	7972348.8300	262733481.900
2019	5 North West	3 TA	Class A Trunk road	299.6000000	186.16	1328652.35800000	9379493.66600	1236229410.0	6077211.6450	241365423.500
2019	8 Yorkshire and The Humber	4 PA	Class A Principal road	3145.0000000	1954.21	48468896.2600000	65874926.58000	7419842901.0	69001521.170	1220136143.000
2019	8 Yorkshire and The Humber	6 MCU	Class C and Unclassified road	26257.0680000	16315.39	285962108.600000	103962407.90000	8375498799.0	102425498.20	1727252125.000
2019	1 South West	1 TM	M or Class A Trunk Motorway	327.9000000	203.75	.00000000	22834924.72000	4638866052.0	21033434.350	8696856280.400
2019	10 West Midlands	6 MCU	Class C and Unclassified road	26716.9510000	16601.14	1701515624.00000	66550770.68000	8265346232.0	64124247.180	1680714611.000
2019	1 South West	5 MB	Class B road	3288.1520000	2043.16	28659955.7200000	51395994.48000	2835552780.0	16221231.730	6664544501.500
2019	6 London	4 PA	Class A Principal road	1747.5000000	1085.85	199046250.300000	269365103.60000	8109482200.0	241064146.80	1641010676.000
2019	7 East of England	1 TM	M or Class A Trunk Motorway	265.0000000	164.66	.00000000	22783138.02000	4440565388.0	19451914.720	1036522779.000
2019	2 East Midlands	1 TM	M or Class A Trunk Motorway	199.9000000	124.21	.00000000	8066516.78800	3141661224.0	11865486.160	748063631.500
2019	9 South East	2 PM	M or Class A Principal Motorway	10.6000000	6.59	.00000000	914310.56010	1362657694.10	504785.04160	19283284.960
2019	6 London	6 MCU	Class C and Unclassified road	12499.2370000	7766.67	183857812.800000	232241999.40000	7420277541.0	31167154.810	1577062030.000
2019	2 East Midlands	4 PA	Class A Principal road	3315.4000000	2060.09	33804969.9700000	63120457.00000	7132604210.0	49948325.760	1257565164.000
2019	4 Wales	5 MB	Class B road	3012.6360000	1871.97	7059972.70900000	16195389.86000	1764267735.0	15049325.370	434096091.200
2019	7 East of England	4 PA	Class A Principal road	3171.7000000	1970.80	50299872.0000000	84520578.93000	8812753991.0	54261816.370	1619017424.000
2019	1 South West	3 TA	Class A Trunk road	747.5000000	464.47	972501.30550000	29111206.97000	3619931494.0	15081721.310	769379181.900
2019	4 Wales	3 TA	Class A Trunk road	1551.4000000	964.00	2960176.71500000	32119478.66000	3633241807.0	22579167.470	760997330.200
2019	10 West Midlands	3 TA	Class A Trunk road	451.6000000	280.61	1608768.16500000	15441299.66000	2351324164.0	8985643.9720	458236225.800
2019	4 Wales	6 MCU	Class C and Unclassified road	26448.7380000	16434.48	92189814.1700000	66439665.15000	4282657741.0	47054453.390	1036621757.000
2019	6 London	5 MB	Class B road	506.2120000	314.55	20803100.9800000	46956689.47000	7804379070.20	101921 60.080	1908122942.700

FIGURE 3.3 Preview of region traffic dataset from the SPSS tool.

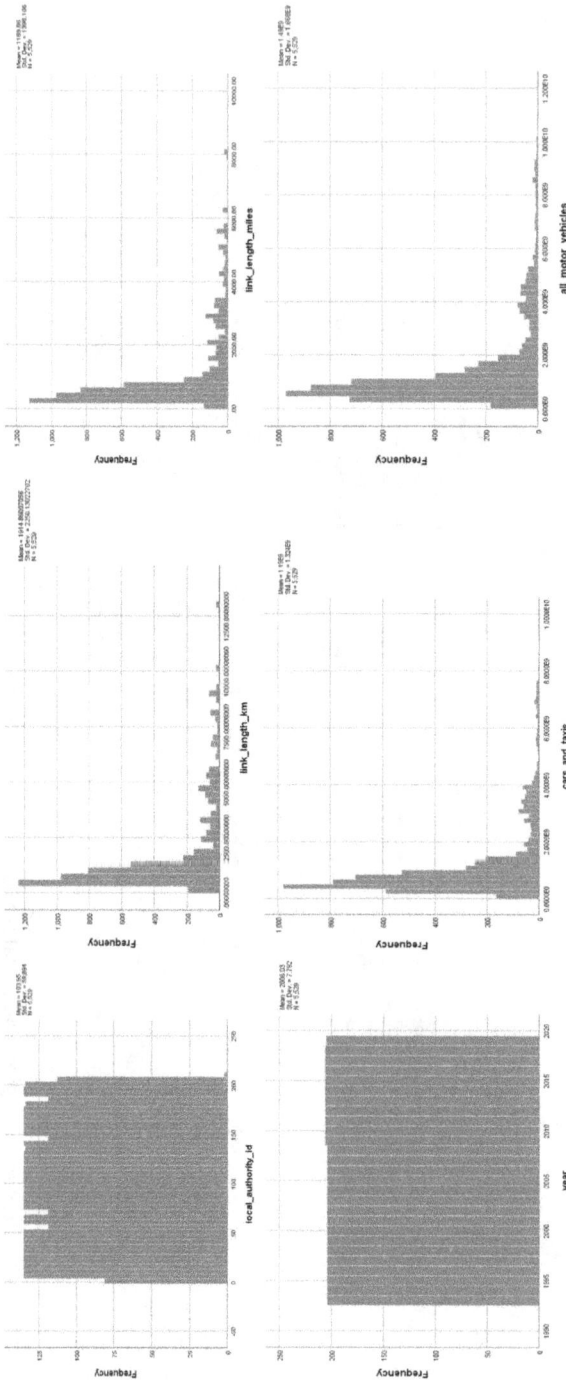

FIGURE 3.4 Histogram graph for "local authority traffic" attributes.

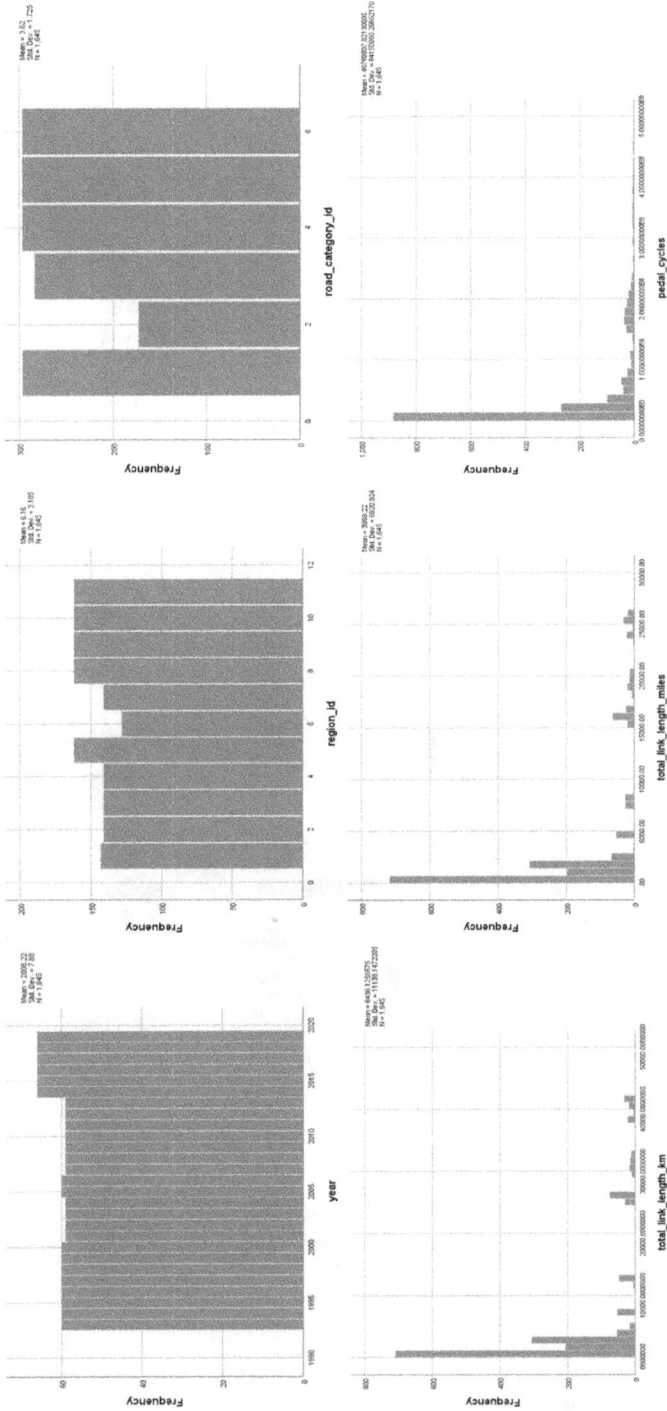

FIGURE 3.5 Histogram graph for "region traffic" attributes.

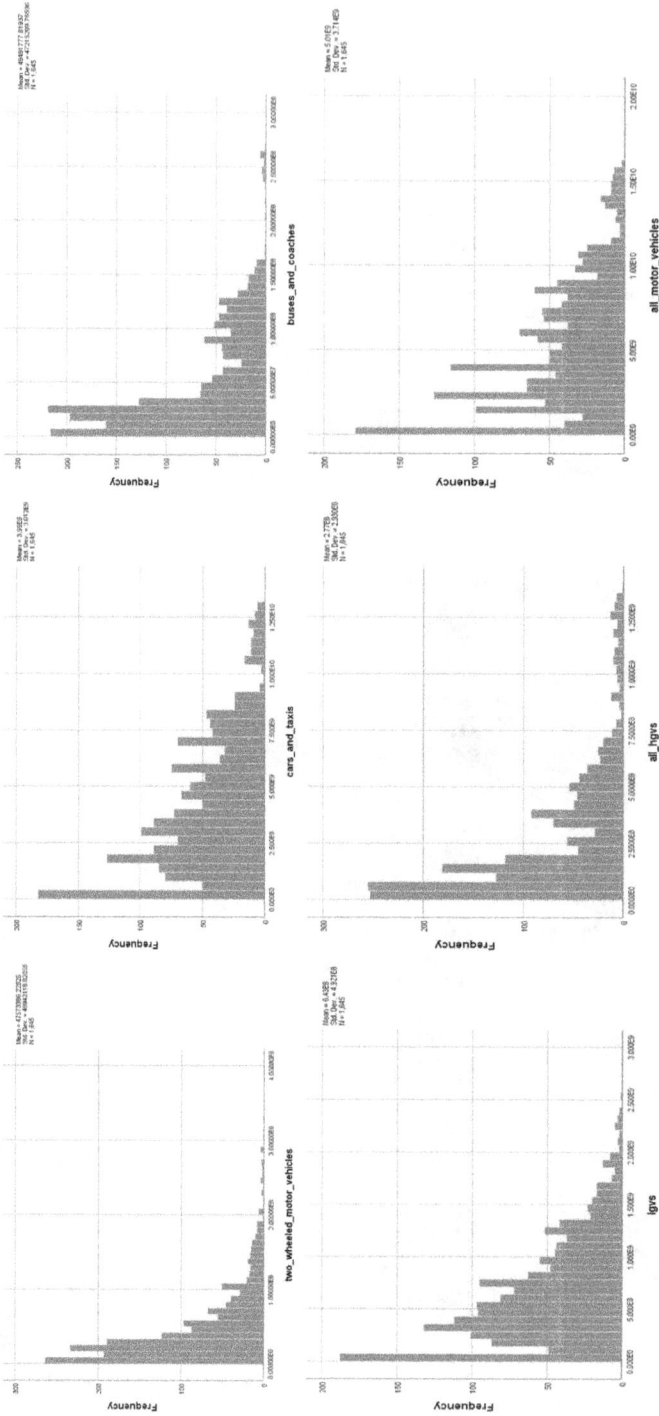

FIGURE 3.6 Histogram graph of "region traffic" for remaining attributes.

The P-P plot is used in the differentiation of symmetrical distributions based on the frequency curve. One of the common features of an equal distribution is that it represents the data by excluding both the smaller and larger values. This is mainly used to visualize the differences in cumulative functions between two distributions. The observed and expected probabilities and observed and expected deviations using P-P plot for "link_length_km", "link_length_miles" attributes of the "local authority traffic" dataset are shown in Figures 3.7 and 3.8.

3.3.2 Machine Learning Models

The process of classifying the various traffic types is known as the categorization of the IBV's network activity information. Machine learning techniques fall into two categories: supervised and unsupervised learning methods. This study employs tree-based supervised ML approaches. The ML step uses a mix of "decision tree" (DT), "random forest" (RF), and "XGBoost" to evaluate and train the traffic analysis using various parameters.

3.3.2.1 Decision Tree (DT)

Information gathering includes the classification method called DT. The classification technique develops a model through inductive reasoning from the collection of pre-classified information. Every data point is defined by the attributes and their values. The categorization may be viewed as a mapping from qualities to classes. Data objects are categorized by DT according to the values of various properties. The decision tree is created using a starting set of previously classified data. In this case, select the characteristics to categorize the classifiers. The numbers of these properties are used to divide up the data pieces.

A selection of data objects is partitioned, and then the procedure is repeated. A subset contains only data elements from the same category. The properties given at every component are separated by DTs. Every node contains a number of links that are each labeled with the characteristic from its parent. The leaf nodes have labels for a number of decision variables to help with classification. Decision trees employ statistical predictors as a classification. It combines a few characteristics through into classification stage and automatically chooses the categories for each piece of data that best describe the target node given all variables.

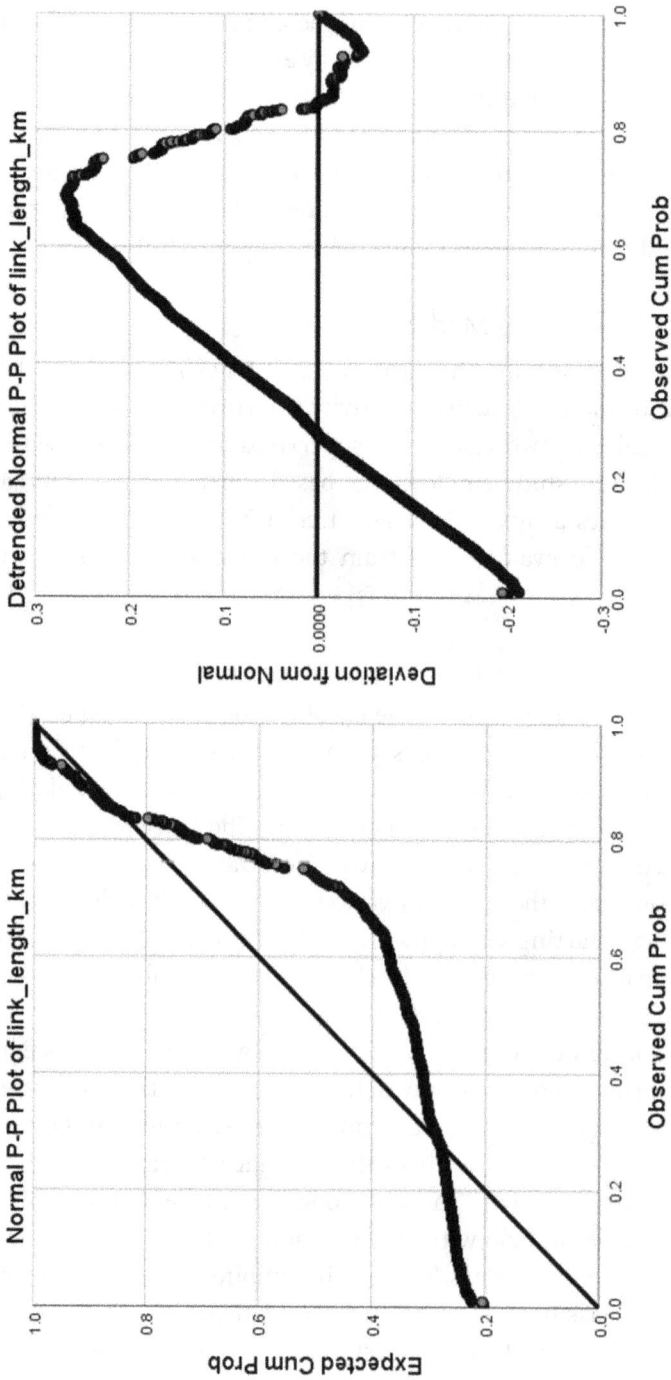

FIGURE 3.7   Observed and expected probabilities for "link_length_km" attribute.

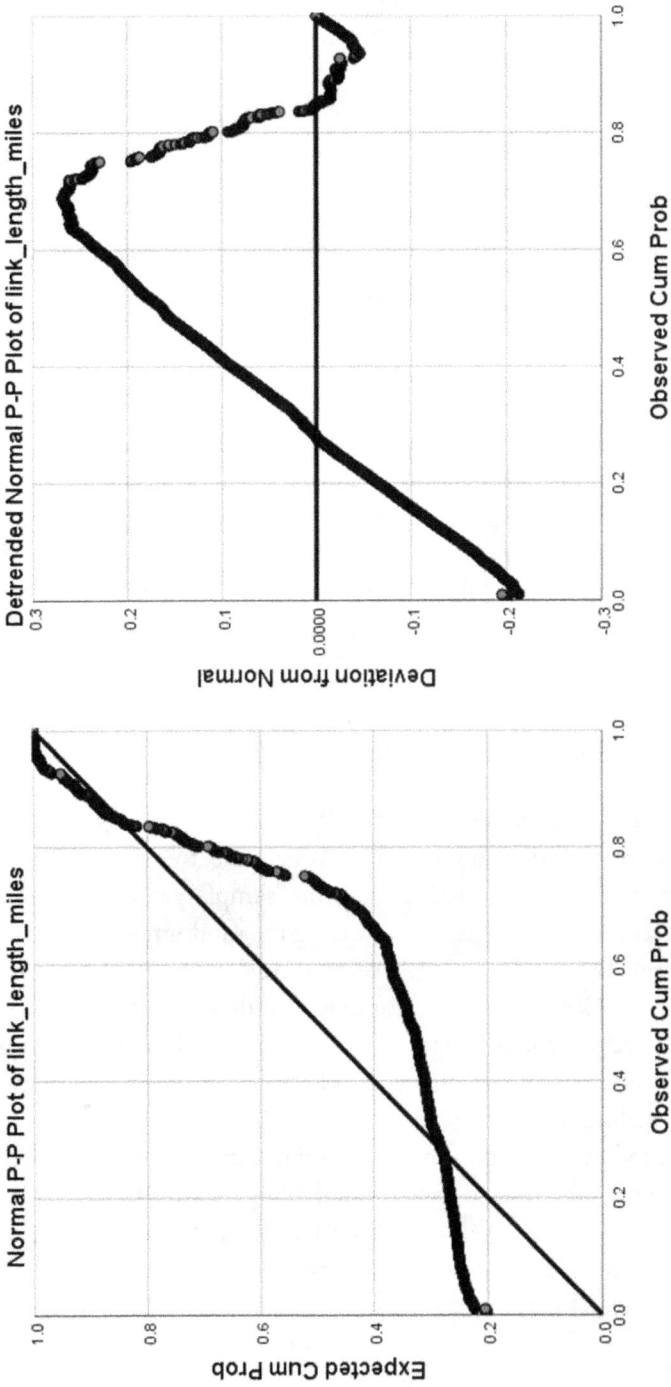

FIGURE 3.8   Observed and expected probabilities for "link_length_miles" attribute.

Let X represent the data point characteristics, and Y represent the category. It bases its choice on the following calculation of the ratio of X to Y, which is calculated by Equation 3.1:

$$DT = \left\{ \left(x_1, x_2, x_3, \ \ldots \ \ldots, \ x_n \right), \left(y_1, y_2, y_3, \ \ldots \ \ldots, y_n \right) \right\} \qquad (3.1)$$

Here $x_1, x_2, x_3, \ \ldots \ \ldots, \ x_n$ represents a feature j of sample i denoted by x_{ij}.

With the use of straightforward classification techniques, a DT attempts to create a classification method that will identify the value of the desired variable.

3.3.2.2 Random Forest (RF)

The output of several DT methods is combined by RF to forecast or categorize the value of a parameter. Depending on an input node of X, which consists of the values of the many evidentiary features assessed for every training range, the RF constructs multiple regression techniques and aggregates the outcomes. After K iterations, the RF predictor of regression for sub-tree $T(x)$ is calculated by Equation 3.2.

$$RF = \frac{1}{K} \sum_1^k T(x) \qquad (3.2)$$

3.3.2.3 The Bagging Procedure Using RF

In order to reduce the connection between the tree branches, a bagging technique generates distinct training sample groups. The bagging approach creates the next subset by resizing the random selection from the entire dataset. The random sample vectors are distributed similarly to the input sequence. Certain input vector may be utilized at least once, whereas other input vectors could be ignored. When exposed to small changes in the input variables, the model becomes better, improving prediction reliability and stability.

From a randomly chosen subset of the data characteristics in the entire information set, RF builds a tree by identifying its most flexible functionality point. As a result, although each tree may be weaker, generalization mistakes can be decreased by lowering the association among trees. Additionally, an RF classification lacks filtering, which makes it operationally inexpensive.

It is important to understand how particular attributes affect the forecasting models in oversampling, where information complexity is significant and also crucial for selecting the most useful evidence. By assessing the precision loss brought on by the fault classification and changing one input evidentiary characteristic while maintaining others fixed, the RF identifies the significance of each variable.

3.3.2.4 XgBoost Classifier

A decision tree strengthening technique is known as boosting tree. A potent technique for enhancing is a boosting tree approach to building on a classifier called as XGBoost X_{gb}. Boosting trees can be thought of as supplementary decision trees. Create a starting lifting tree, and Equation 3.3 can be used to build the explanatory modeling for solving tree t.

$$X_{gb} = \sum_1^t f_k(x_i) \tag{3.3}$$

When resolving classification trees, use Equation 3.4 to minimize the objective function. By evaluating every feasible separating point for each characteristic, the best dividing parameter and dividing point are determined by Equation 3.4.

$$Objective\ function = \sum_i l(y_i', y_i) + \sum_k f_k \tag{3.4}$$

Here, $l(y_i', y_i)$ represents loss function and f_k determines regulatory function.

3.3.2.5 Support Vector Machine (SVM)

The choice of a kernel function and its coefficients is essential because a solid kernel generates a representation that closely matches the structure underlying the source data. Using parameters in SVM, the information is moved into a higher-dimensional region. RBF kernels are the best choice for a speedy and reliable process of learning since they do complex and non-linear transformations more efficiently than other kernels and have straightforward feature computation. Kernel characteristics could be added and modified to specify the kernel functions. For the SVM to operate well, these model parameters must be chosen carefully. Two factors that are associated with this kernel must also be properly taken into consideration when retraining an RBF-SVM. Gamma " 'γ'" and penalty factors are the two important parameters connected to the kernel function. The penalty variable indicates how the kernel influences the network and instructs the SVM optimization on how much error to minimize for each training instance. Applicable to everyone's SVM kernels, this compromises the misinterpretation of training data for the selection surface simplification.

3.3.2.6 Principal Component Analysis (PCA)

A frequency distribution indicates the eigenvalues of the variables in a data transformation. The statistical model helps determine how many principal components to keep while completing a PCA. The objective of this approach is to find statistically significant factors. The degree of variance that each prediction represents is used to order the indicated variations for every classifier. The dataset feature variations are ordered from smallest to highest, with the diagonal elements following suit. This demonstrates that perhaps the eigenvalue, which accounts for the majority of the information, is available variation is represented by the final components.

3.3.3 Experimental Analysis

3.3.3.1 Receiver Operating Characteristic (ROC) Curve

ROC is a graphical representation showing the classification threshold values based on the performance of classification models. This curve includes two parameters, namely "True Positive Rate" (TPR) and "False Positive Rate" (FPR), where "Recall" is the general notation of "True Positive" rate, and "False Positive" rate is given by sum of false positives and true negatives to the total number of false positives.

The ROC for the "local authority traffic" dataset with respect to the year 2019 is shown in Figure 3.9 and for the year 2018 in Figure 3.10 as test values.

The ROC for the "region traffic" dataset with respect to the year as a positive actual state in 2019 is shown in Figure 3.11 and for the year 2018 in Figure 3.12.

3.3.3.2 Area Under Curve (AUC)

The AUC is a measure among two-dimensional areas representing the performance among all possible threshold values. It represents the probability among TPR and FPR at different intervals of threshold values and necessarily divides the signal from its noise. This provides an aggregate measure while computing binary classifiers among different threshold values.

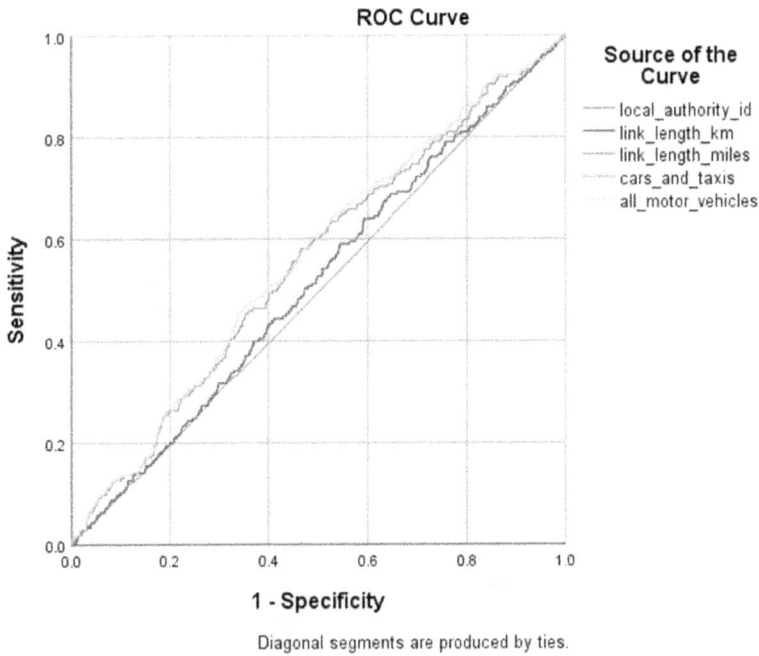

FIGURE 3.9   ROC for "local authority traffic" dataset with test value for 2019.

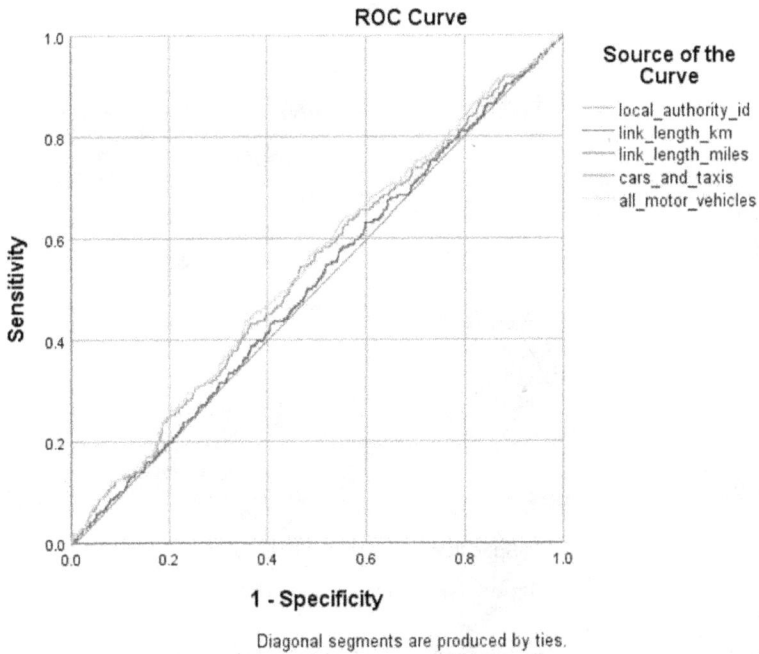

FIGURE 3.10   ROC for "local authority traffic" dataset with test value for 2018.

ROC Curve

Source of the Curve
- region_id
- road_category_id
- total_link_length_km
- total_link_length_miles
- pedal_cycles
- two_wheeled_motor_vehicles
- cars_and_taxis
- buses_and_coaches
- lgvs
- all_hgvs
- all_motor_vehicles

Diagonal segments are produced by ties.

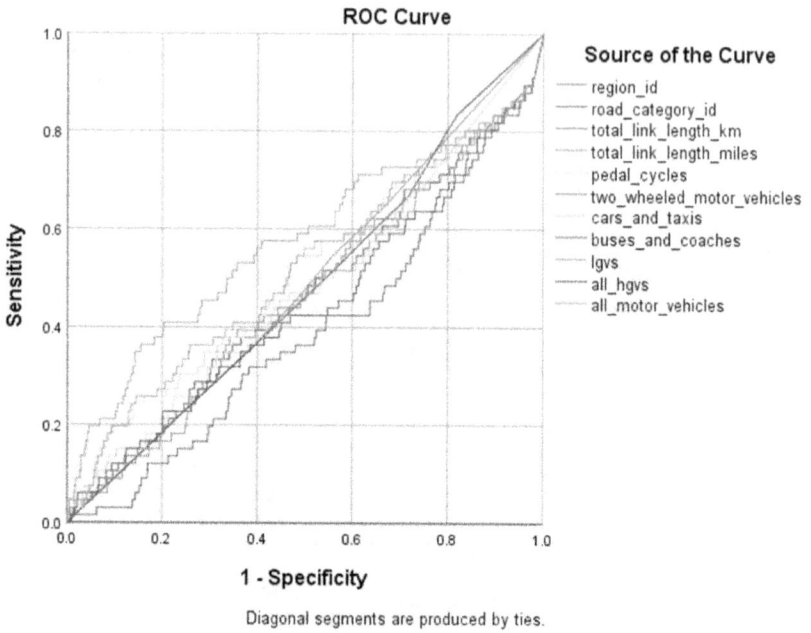

FIGURE 3.11   ROC for "region traffic" dataset with test value 2019.

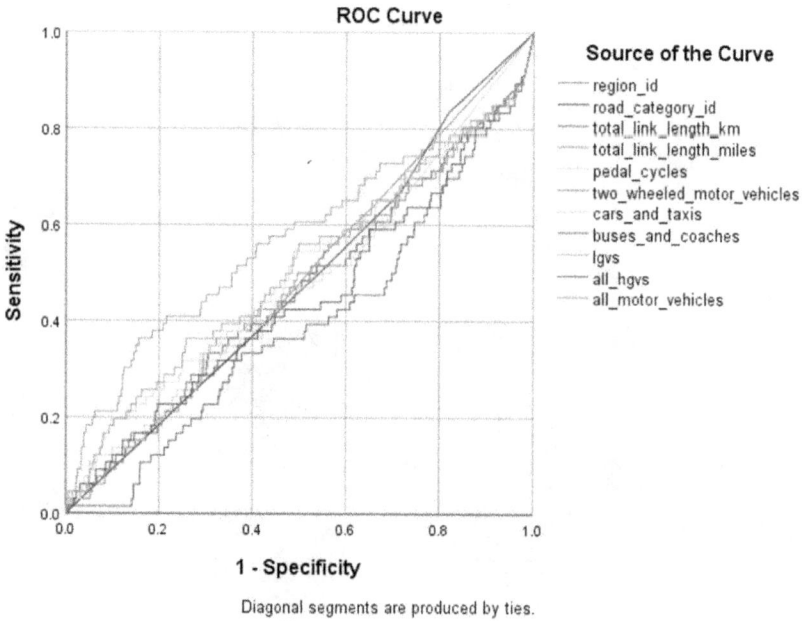

ROC Curve

Source of the Curve
- region_id
- road_category_id
- total_link_length_km
- total_link_length_miles
- pedal_cycles
- two_wheeled_motor_vehicles
- cars_and_taxis
- buses_and_coaches
- lgvs
- all_hgvs
- all_motor_vehicles

Diagonal segments are produced by ties.

FIGURE 3.12   ROC for "region traffic" dataset with test value 2018.

TABLE 3.4   AUC Values for Figure 3.9

Area Under the Curve	
Test Result Variable(s)	**Area**
local_authority_id	.498
link_length_km	.513
link_length_miles	.513
cars_and_taxis	.554
all_motor_vehicles	.561

TABLE 3.5   AUC Values for Figure 3.10

Area Under the Curve	
Test Result Variable(s)	**Area**
local_authority_id	.498
link_length_km	.508
link_length_miles	.508
cars_and_taxis	.534
all_motor_vehicles	.543

TABLE 3.6   AUC Values for Figure 3.11

Area Under the Curve	
Test Result Variable(s)	**Area**
region_id	.485
road_category_id	.480
total_link_length_km	.465
total_link_length_miles	.465
pedal_cycles	.479
two_wheeled_motor_vehicles	.463
cars_and_taxis	.507
buses_and_coaches	.390
Lgvs	.574
all_hgvs	.443
all_motor_vehicles	.515

The AUC values for Figures 3.9 and 3.10 of the "local authority traffic" dataset is shown in Tables 3.4 and 3.5.

The AUC values for Figures 3.11 and 3.12 of the "region traffic" dataset are shown in Tables 3.6 and 3.7.

TABLE 3.7   AUC Values for Figure 3.12

Area Under the Curve	
Test Result Variable(s)	**Area**
region_id	.485
road_category_id	.480
total_link_length_km	.464
total_link_length_miles	.464
pedal_cycles	.475
two_wheeled_motor_vehicles	.471
cars_and_taxis	.499
buses_and_coaches	.392
Lgvs	.567
all_hgvs	.443
all_motor_vehicles	.508

3.4 CASE STUDY AND APPLICATION

3.4.1 Traffic Light Management System

Different modes of transportation or other congested road objects, including individuals and small transportation devices, are also included in traffic signal control. These elements are also taken into account in some research. On RL-based modeling and vehicle traffic taken into account depending on such unrestricted turn boundaries, each turning for a route has a different learning impact. Pedestrians might also have an effect on traffic movement because junctions occasionally have the capability to interrupt stated policy. When they use an input item, such as a relocation demand button, to achieve that, the junction allows them the appropriate amount of time to travel.

3.4.2 Reinforcement Learning (RL)

As was previously noted, the three major components of reinforcement learning are situation, activity, and reward. The traffic signal layout choices allow for the right combination and alignment of these components. Rather than mixing numerous characteristics, one variable from each of these criteria is typically employed in the formulation of the state or incentive.

3.4.2.1 Single-Agent RL

The study could potentially propose a wide categorization of model-free classifiers by dividing them into two different sets, value and policy based, while analyzing the studies that utilize RL to solve EV routing challenges. Whereas deep Q-network (DQN) is the foundation of the deep-RL technique that estimates the action and responsive function using deep neural networks, Q-learning is the conventional RL approach for training action value functionality in the value-based set.

3.4.2.2 Multi-agent RL

Advanced models in RL necessitate the use of several agents to acquire and perform various tasks concurrently, converting single-agent learning algorithms to multi-agent evolutionary algorithms. The agents in the Markov event, in contrast to the multi-agent and single-agent methods, are associated with one another in the context and have the ability to affect both the environmental behavior and the best outcomes. It is harder to learn the best policies for each agent and control how they connect with one another because the complexities of the surroundings implied define the agent policies, and because these are constantly being changed during learning, instability problems are easily initiated.

3.4.2.3 Q-Learning

A tabular method that utilizes temporal differential learning is known as Q-Learning. Assuming discontinuous spaces, including both action and state, the best Q-value feature $Q^*(S,a)$ can be seen as a 4×3 table. The most effective action to take after assessing the current situation is defined in Equation 3.5.

$$a_t = argmax Q^*\left(S_t, a\right)$$ (3.5)

It involves deciding the row's highest Q-value, which corresponds to state S_t, and providing the action relating to that Q-value. A Q-table is employed in Q-learning to estimate the Q-value equation. It modifies each component of a table at a moment after initializing each of them to zero initially.

3.4.2.4 Deep Q-Learning Network (DQLN)

DQLN uses a deep neural system as a portion of the available function to describe the Q-value variable in multi-layered continuous action in order to get around the decision variable space in Q-learning. The usage of an event replayed buffer, consisting of hold tuples that the user communicates with the context, is another essential method for stabilization. The mapping function is continual; hence, it recognizes the existence of a relationship between two successive events. Because the training dataset is also not consistent, the strategy is easily overfitted when a group of observations is used immediately as the classifier is trained. This issue can be resolved by batching a limited number of instances taken at irregular intervals from the buffer. This not only guarantees that the training sets are diverse and distributed evenly but also decreases the size of every group to a level that speeds up learning.

3.4.3 Fusion-Based STCS

Smart cities are using an intelligent STCS to track and manage congestion problems using machine learning methods. In this method, IBVs are capable devices that are used to gather traffic statistics. With the use of this approach, data can be maintained and alerts sent from one road point to another point. Sensor information is subsequently sent to the sensory layers, where it is processed through the processing, learning, performing, and testing phases. V-to-I communication and V-to-V broadcasting are made possible by the vehicle network. Therefore, through the use of this network, the acquisition of sensory information via wireless transmission vehicles rather than expensive cellular connectivity is made possible. First, information can be compiled from ground stations along the sides of the road using V-to-V or V-to-I communication. Further, to track and redirect traffic congestion, the ground station server can send the information it has gathered straight to the server.

3.4.4 Intelligent Transportation Systems (ITS)

The term "Intelligent Transportation Systems (ITS)" refers to a group of innovations and technologies that have been widely embraced and are now used to equip transportation and maintenance vehicles with communication and data equipment. By sensibly combining technological advances for evaluation, detection, preparation, and evaluation with information systems for simulated and real-time regulation, networks, and modern communications, ITS addresses the requirement to measure, monitor, and resolve issues relating to traffic jams. Sensor and reactive cameras form the foundation of the current generation of congestion surveillance technology. This new technology of detectors was initially deployed as an element of the active monitoring of national highways and underpasses, and it certainly proved to be more efficient, practical, and suitable.

3.4.5 Surveillance Monitoring System (SMS)

The phone camera's function is to capture objects in a certain location while also comprehending and differentiating vehicles so that it may derive the related license plate from such a picture using pre-processing methods like slicing, resizing, and zooming. Once the plate-related image has been acquired, it is fed into a deep network that has been specially trained to identify the characters and digits on the license plate, and the information is then turned into a phrase. For each recording location, image processing techniques are put into place. A sequence of procedures is run by the program to identify the license plate. This method is in the process of searching for it and identifying the license plate in the image, as well as the size and position of the plate. This corrects the frame's angle and modifies the measurements to the appropriate dimensions.

By changing from RGB to grayscale images, for example, some methods of image processing can be used in the form that is needed. It is utilized to widen the gap between both the characters and the receiver for the license plate. It is possible to reduce the optical noise in the image using a Gaussian filter.

This technique segments the symbols on the plate based on how they are found individually. The digital translation of written or visual information into machine code is known as character recognition. Analysis of identities and roles against regional regulations provides a more secure and dependable result by averaging the recognized value across several fields, particularly since each and every image could contain certain light that is reflecting or that is partially hidden.

3.4.6 Clustering-Based Vehicular Networks

To manage nodes in an orderly manner, clustering is the method of creating a group of networks with similar characteristics depending on certain predetermined parameters. In a systematic way, it simplifies transportation. By removing the hidden node issue, forming reasonable categories based on closeness, and maximizing network bandwidth using bandwidth, VANET also takes advantage of clustering advantages. Vehicles within a network, for example, don't need to be aware of the routing table from outside the cluster. Durability is provided by clustering in networking. Several studies have looked into network technologies and approaches for mobile ad hoc networks to this point.

3.4.7 Network Anomaly Detection Using Image Processing

Data visualization has the advantage over standard text data in that within the image, pixel feature extraction can be used to depict the information in an appropriate data format. Using image processing techniques, the detection system would place emphasis on created images during visualization rather than hidden in terms of static or dynamic aspects. In order to solve the issue of a single detection device's minimal exposure, to create grayscale features in images from network traffic information so that features can be analyzed, and to improve the evaluation of data transmitted on a global scale, this approach compiles the evaluation findings from various detection methods into a spatial context for analysis. Using a classification method in combination with solution space, information is mapped to a 3D visual area for evaluation. This enables the detection of unusual communication networks. Data feature extraction can help increase the identification reliability and stability of the image processing model to some level by analyzing the objects on overall properties.

3.4.8 Deep Learning for Vehicular Detection Networks

Deep learning, a branch of machine learning, is distinguished by its capacity to imitate neural network-based training like a human brain. As a result, it has recently demonstrated strong performance in applications for vision, text, image, and language processing. Many studies suggest that a deep learning-based vehicle detection model can outperform the old approach because of deep learning's superior learning abilities. A multi-layered learning and development strategy for both temporal and spatial aspects of data from network traffic is provided, which lowers the false and positive detection rates for models by combining convolutional neural networks (CNN) with long- and short-term memory networks (LSTM). These architectures contain the responsive optimization technique for particle swarms to automatically detect the fully connected layers' characteristics in order to improve the effectiveness of vehicle detection. The benefit of these techniques is that they are able to significantly enhance the precision and effectiveness of identification in the particular instance of a large volume of information. But their drawback is that they are capable of only recognizing trained data. So, the outcomes of the identification could still be uncertain whenever the recognition connectivity is disrupted by different attacks.

3.4.9 Internet-of-Things (IoT)

IoT is the networking and interconnecting of trillions of different objects to develop a smart community. These systems use established communication systems to communicate and share data across various applications. To create and estimate IoT network, three layers comprise its DL model. The forecast of internet quality performance is addressed in the first layer. Evaluating the performance of many other IoT environments is the responsibility of the second layer. Predicting the effectiveness of the IoT architecture is the major goal of the third layer. The problem of renewable energy in public organizations and the development of ML algorithms for energy usage estimation. The design of an intelligent ML-based energy control system for public organizations can be used as a component of the smart traffic system. The data is taken from a centralized repository of public areas that was created via an open smart IoT network as well as an energy information system for management.

3.4.10 Cloud Computing

The term "cloud computing" refers to a technology that enables users to access information anywhere at any time from any place. It consists of several data centers to which a huge number of users around the world can connect via the internet. In order to progress urban populations, several cities have adopted cloud-based technologies for traffic control, public services, urban security, and so on. As complex structures, smart cities need more than just planning, creation, execution, and surveillance. Additionally, for monitoring the entire development process of smart cities and verifying the futuristic and effectiveness of its development, the main strategy for creating smart traffic public infrastructure is to use a cloud infrastructure.

3.4.11 Edge Computing

Edge computing is promoted as a way to reduce and neutralize the drawbacks of cloud services. By processing data as near to the origin as possible, this is a technology and computing method that enhances cloud services. Because enormous volumes of information are handled there before being transmitted to the cloud core systems, the edges have a sizable compute area. The edge is a decentralized platform for communications and processing that can be applied to a variety of situations, such as smart transportation systems, intellectual aids, and medical and social services. For time-sensitive activities, the edge reduces edge delay and enhances the performance of certain programs.

3.4.12 Fog Computing

The use of cloud computing is encouraged as an option for performing all data analysis online. IoT devices can be used with fog computing, a sort of cloud technology that reaches the edge of a network. Fog, on the contrary side, puts data processing and analysis components and users nearer together. In terms of the concept of fog computing, it consists of a vast volume of massive, wide and varied, wireless, occasionally impartial, prevalent, and decentralized facilities that communicate and may even collaborate together with the network to perform storage and computation tasks without the assistance of outside parties.

Fog and IoT are examples of new enabling developments that can be used to make it easier to create smart traffic management, which is advantageous for the development of urban cities, horticulture, other sectors such as tourism, and transportation management. Therefore, establishing smart traffic will considerably strengthen the area's capacity for continuous expansion.

3.5 CONCLUSIONS WITH FUTURE RESEARCH SCOPES

Using real-time information, the STCS is improved by identifying the required modes of transportation. Reducing the overcrowding traffic by maintaining the traffic signals at junctions is the major factor. Gathering public attention is an important part of avoiding traffic jams. Because of overcrowded population in smart cities, there is a high chance of common traffic. Such situations can't be controlled and handled manually. Therefore, controlling the traffic system using machine learning is an essential mechanism in smart cities. In this study, various ML models are discussed for handling the traffic data. The benefits of STCS show the impact of automated controlling system, which is necessary in present-day situations. To understand the statistical analysis, we have used "local authority traffic" and "region traffic" datasets to understand the congestion situations. For this analysis, we have used P-test analysis and AUC values and ROC curves to identify the significance of traffic data. In the future, there is a chance to perform all experimental results using ML algorithms by using all the STCS applications discussed in this study.

REFERENCES

1. Anyanwu, G. O., Nwakanma, C. I., Lee, J. M., & Kim, D. S. (2023). RBF-SVM kernel-based model for detecting DDoS attacks in SDN integrated vehicular network. *Ad Hoc Networks*, 140, 103026. https://doi.org/10.1016/j.adhoc.2022.103026

2. Li, C., Zheng, P., Yin, Y., Wang, B., & Wang, L. (2023). Deep reinforcement learning in smart manufacturing: A review and prospects. *CIRP Journal of Manufacturing Science and Technology*, 40, 75–101. https://doi.org/10.1016/j.cirpj.2022.11.003

3. Mai, J., Wu, Y., Liu, Z., Guo, J., Ying, Z., Chen, X., & Cui, S. (2023). Anomaly detection method for vehicular network based on collaborative deep support vector data description. *Physical Communication*, 56, 101940. https://doi.org/10.1016/j.phycom.2022.101940

4. Heidari, A., Navimipour, N. J., & Unal, M. (2022). Applications of ML/DL in the management of smart cities and societies based on new trends in information technologies: A systematic literature review. *Sustainable Cities and Society*, 104089. https://doi.org/10.1016/j.scs.2022.104089

5. Anedda, M., Fadda, M., Girau, R., Pau, G., & Giusto, D. (2023). A social smart city for public and private mobility: A real case study. *Computer Networks*, 220, 109464. https://doi.org/10.1016/j.comnet.2022.109464

6. Boualouache, A., & Engel, T. (2022). A survey on machine learning-based misbehavior detection systems for 5g and beyond vehicular networks. *arXiv Preprint*. https://doi.org/10.48550/arXiv.2201.10500

7. Alladi, T., Chamola, V., Sahu, N., Venkatesh, V., Goyal, A., & Guizani, M. (2022). A Comprehensive survey on the applications of blockchain for securing vehicular networks. *IEEE Communications Surveys & Tutorials*. https://doi.org/10.48550/arXiv.2201.04803

8. Khan, A. S., Sattar, M. A., Nisar, K., Ibrahim, A. A. A., Annuar, N. B., Abdullah, J. B., & Karim Memon, S. (2023). A survey on 6G enabled light weight authentication protocol for UAVs, security, open research issues and future directions. *Applied Sciences*, 13(1), 277. https://doi.org/10.3390/app13010277

9. Alkinani, M. H., Almazroi, A. A., Adhikari, M., & Menon, V. G. (2022). Artificial intelligence-empowered logistic traffic management system using empirical intelligent XGBoost technique in vehicular edge networks. *IEEE Transactions on Intelligent Transportation Systems*. https://doi.org/10.1109/TITS.2022.3145403

10. Ahmed, U., Lin, J. C. W., Srivastava, G., Yun, U., & Singh, A. K. (2022). Deep active learning intrusion detection and load balancing in software-defined vehicular networks. *IEEE Transactions on Intelligent Transportation Systems*. https://doi.org/10.1109/TITS.2022.3166864

11. Song, W., Rajak, S., Dang, S., Liu, R., Li, J., & Chinnadurai, S. (2022). Deep learning enabled IRS for 6g intelligent transportation systems: A comprehensive study. *IEEE Transactions on Intelligent Transportation Systems*. https://doi.org/10.1109/TITS.2022.3184314

12. Tomar, I., Sreedevi, I., & Pandey, N. (2022). State-of-art review of traffic light synchronization for intelligent vehicles: Current status, challenges, and emerging trends. *Electronics*, 11(3), 465. https://doi.org/10.3390/electronics11030465

13. Zheng, Z., & Bashir, A. K. (2022). Graph-enabled intelligent vehicular network data processing. *IEEE Transactions on Intelligent Transportation Systems*, 23(5), 4726–4735. https://doi.org/10.1109/TITS.2022.3158045

14. Elfatih, N. M., Hasan, M. K., Kamal, Z., Gupta, D., Saeed, R. A., Ali, E. S., & Hosain, M. S. (2022). Internet of vehicle's resource management in 5G networks using AI technologies: Current status and trends. *IET Communications*, 16(5), 400–420. https://doi.org/10.1049/cmu2.12315

15. Aldhyani, T. H., & Alkahtani, H. (2022). Attacks to automatous vehicles: A deep learning algorithm for cybersecurity. *Sensors*, 22(1), 360. https://doi.org/10.3390/s22010360

16. Khasawneh, A. M., Helou, M. A., Khatri, A., Aggarwal, G., Kaiwartya, O., Altalhi, M., . . . AlShboul, R. (2022). Service-centric heterogeneous vehicular network modeling for connected traffic environments. *Sensors*, 22(3), 1247. https://doi.org/10.3390/s22031247

17. Javed, A. R., Hassan, M. A., Shahzad, F., Ahmed, W., Singh, S., Baker, T., & Gadekallu, T. R. (2022). Integration of blockchain technology and federated learning in vehicular (IOT) networks: A comprehensive survey. *Sensors*, 22(12), 4394. https://doi.org/10.3390/s22124394

18. Wahab, O. A., Mourad, A., Otrok, H., & Bentahar, J. (2016). CEAP: SVM-based intelligent detection model for clustered vehicular ad hoc networks. *Expert Systems with Applications*, 50, 40–54. https://doi.org/10.1016/j.eswa.2015.12.006

19. Rani, P., & Sharma, R. (2023). Intelligent transportation system for internet of vehicles based vehicular networks for smart cities. *Computers and Electrical Engineering*, 105, 108543. https://doi.org/10.1016/j.compeleceng.2022.108543

20. Saleem, M., Abbas, S., Ghazal, T. M., Khan, M. A., Sahawneh, N., & Ahmad, M. (2022). Smart cities: Fusion-based intelligent traffic congestion control system for vehicular networks using machine learning techniques. *Egyptian Informatics Journal*. https://doi.org/10.1016/j.eij.2022.03.003

21. Kiani, F., & Saraç, Ö. F. (2023). A novel intelligent traffic recovery model for emergency vehicles based on context-aware reinforcement learning. *Information Sciences*, 619, 288–309. https://doi.org/10.1016/j.ins.2022.11.057

22. Qiu, D., Wang, Y., Hua, W., & Strbac, G. (2023). Reinforcement learning for electric vehicle applications in power systems: A critical review. *Renewable and Sustainable Energy Reviews*, 173, 113052. https://doi.org/10.1016/j.rser.2022.113052

23. Ma, D., Wang, P., Song, L., Chen, W., Ma, L., Xu, M., & Cui, L. (2023). A lightweight deployment of TD routing based on SD-WANs. *Computer Networks*, 220, 109486. https://doi.org/10.1016/j.comnet.2022.109486

24. Ayyub, M., Oracevic, A., Hussain, R., Khan, A. A., & Zhang, Z. (2022). A comprehensive survey on clustering in vehicular networks: Current solutions and future challenges. *Ad Hoc Networks*, 124, 102729. https://doi.org/10.1016/j.adhoc.2021.102729

25. Park, S., Yoo, Y., & Pyo, C. W. (2022). Applying DQN solutions in fog-based vehicular networks: Scheduling, caching, and collision control. *Vehicular Communications*, 33, 100397. https://doi.org/10.1016/j.vehcom.2021.100397

26. Wang, R., Jiang, X., Zhou, Y., Li, Z., Wu, D., Tang, T., . . . Badenko, V. (2022). Multi-agent reinforcement learning for edge information sharing in vehicular networks. *Digital Communications and Networks*, 8(3), 267–277. https://doi.org/10.1016/j.dcan.2021.08.006

27. Elbir, A. M., Soner, B., Çöleri, S., Gündüz, D., & Bennis, M. (2022, September). Federated learning in vehicular networks. In *2022 IEEE International Mediterranean Conference on Communications and Networking (MeditCom)* (pp. 72–77). IEEE.

28. Park, H. (2022). Edge based lightweight authentication architecture using deep learning for vehicular networks. *Journal of Internet Technology*, 23(1), 193–200.

29. Elsagheer Mohamed, S. A., & AlShalfan, K. A. (2021). Intelligent traffic management system based on the internet of vehicles (IoV). *Journal of Advanced Transportation*, 2021. https://doi.org/10.1155/2021/4037533

30. Humayun, M., Afsar, S., Almufareh, M. F., Jhanjhi, N. Z., & AlSuwailem, M. (2022). Smart traffic management system for metropolitan cities of kingdom using cutting edge technologies. *Journal of Advanced Transportation*, 2022. https://doi.org/10.1155/2022/4687319

31. Hilmani, A., Maizate, A., & Hassouni, L. (2020). Automated real-time intelligent traffic control system for smart cities using wireless sensor networks. *Wireless Communications and Mobile Computing*, 2020. https://doi.org/10.1155/2020/8841893

32. Alvi, A. N., Javed, M. A., Hasanat, M. H. A., Khan, M. B., Saudagar, A. K. J., Alkhathami, M., & Farooq, U. (2022). Intelligent task offloading in fog computing based vehicular networks. *Applied Sciences*, 12(9), 4521. https://doi.org/10.3390/app12094521

33. www.kaggle.com/arashnic/road-trafic-dataset?select=region_traffic.csv

CHAPTER 4

Revolutionizing Transportation

The Power of Artificial Intelligence

Pooja Sharma

4.1 INTRODUCTION

Technology has become an indispensable part of every sector worldwide, including banking, healthcare, education, retail, insurance, farming, social media, and sports industries. It has significantly transformed the way businesses function, resulting in reduced costs and increased efficiency. The transport industry is one of the crucial areas where technology can play a crucial role in overcoming several challenges such as traffic congestion, routing issues, unexpected delays, road accidents, and poor logistics resulting in financial losses.

The transport industry plays a pivotal role in the movement of people and goods across different geographical locations, particularly to remote areas. The efficient operation of courier services is highly dependent on a well-established transport system that delivers goods and products to the right place at the right time. To automate and streamline these operations, governments and organizations have incorporated advanced technologies such as machine learning (ML), artificial intelligence (AI), and the Internet of Things (IoT).

4.2 ARTIFICIAL INTELLIGENCE

AI is a broad field of computer science focused on creating machines that can perform cognitive functions like the human brain but at a faster rate. The term AI was first coined by John McCarthy in 1956, and in more than

DOI: 10.1201/9781003409502-4

60 years since, AI has become increasingly popular thanks to the abundance of available data. By training AI applications on large volumes of data, complex tasks can be processed with ease, leading to more effective decision-making and process automation [1]. AI can also offer solutions to global challenges like climate change, water scarcity, and the creation of sustainable, smart cities in developed nations [2].

The United States and China currently lead the world in AI, with other countries such as Canada, the UK, Russia, and France also making significant investments in the field [3]. India is expected to become a major player in AI in the coming years, thanks to factors like its abundant data, demographics, and structural advantages. The Indian government has invested $477 million in the Digital India project, which aims to develop AI, machine learning, IoT, big data, and related technologies [4]. One area where AI has been particularly successful in India is crowd and traffic management.

Cities around the world are struggling with issues like traffic congestion, logistics, and transportation due to rapidly growing populations and increased numbers of vehicles on the roads. AI is emerging as a promising solution to these problems and has been recognized as such by the World Economic Forum [5]. Techniques like artificial neural networks, fuzzy logic models, genetic algorithms, ant colony optimization, and simulated annealing can help address issues like traffic congestion, improve travel time reliability, routing, and overall productivity and economics [6].

4.3 SMART TRANSPORTATION

In recent years, there has been an exponential increase in the amount of data generated, including text, images, audio, and videos, due to the availability of various technologies and internet access. This data is used to make informed decisions in businesses, in government sectors, and for the benefit of society. Similarly, the transport industry plays a significant role in the urbanization of cities and countries, as it directly impacts people, processes, and profits.

To develop smart transport, automobile companies have built devices that can be implanted directly into vehicles used for transporting products and the public. Data generated through these devices is remotely monitored by experts, and by processing it using AI algorithms, governments and businesses can make real-time decisions. The Transport Management System (TMS) deals with transport operations and is growing rapidly, with objectives including effective route planning, load optimization, and improved logistics and transparency [7].

This chapter focuses on the smart transport system, which is a part of TMS. It collects and summarizes data from research papers, government reports, and journal articles to provide a comprehensive understanding of the subject. This study will help businesses and government departments to develop AI-based smart transport systems and provide better solutions.

4.4 LITERATURE SURVEY

As previously mentioned, AI has made significant contributions in various industries, including transportation. This section will highlight some of the notable ways AI has impacted the transport industry. AI has proven to be a valuable tool in addressing issues related to the design, time scheduling, operation, and administration of logistical systems [8]. By using AI algorithms, it is possible to better understand the complex relationships among various features of transportation systems [1]. Experts in public transport expect that AI will play a crucial role in the future of mobility [9]. Smart surveillance control systems, management of control vehicles, and routing are discussed in [10], while [11] explores real-time traffic management by comparing two integrated autonomous agents.

Autonomous vehicles could help address major transport issues such as safety, traffic control, increased mobility needs, and pollution control, as outlined in [12]. Furthermore, autonomous vehicles could have a significant impact not just on individual industries but on a country's or nation's economy as a whole [13]. New models and forms of manufacturing vehicles are proposed in [14] with the aim of creating smart vehicles. The logistics of transportation systems are addressed in [15], as they connect various logistic activities.

4.5 DESIGN STUDY

The previous studies reveal that although AI-powered intelligent transport systems have been developed, there is still room for improvement and further research. This chapter aims to explore the implementation of smart transport in various countries. Transportation plays a crucial role in any country's economy, but it often faces challenges that can impede progress at the local, regional, and national levels. Smart transport solutions employ advanced technologies to address these challenges, enabling

TABLE 4.1   Sub-systems of Smart Transport System

Sub Systems of Smart Transport	Role
Smart traffic management system	Real-time congestion and road management
Smart safety management system	Safeguarding people, goods, and vehicles on road
Smart public transport system	Defining optimal routes for transporting public through various transport media
Smart manufacturing and logistics system	Integrating advanced technology in manufacturing of automobiles along with fast and safe delivery of goods

timely delivery of goods and meeting customer demands while reducing the costs associated with supply chain management. Remote collection and analysis of data enables real-time monitoring of deliveries through dashboards, promoting transparency in the system. These smart transport applications and facilities are utilized by logistics companies, manufacturers, and distributors.

Smart transportation systems leverage AI and ML to make intelligent decisions, make accurate predictions, and conduct risk analysis. Popular technologies employed include IoT, ML, and computer vision. Previous research suggests that a holistic approach to smart transportation should include both transport infrastructure and management. Smart transportation comprises several sub-systems, such as smart traffic management, smart safety management, smart public transport, and smart manufacturing and logistics systems. Table 4.1 summarizes the role of each of these subsystems, highlighting the benefits of AI in building smart transportation systems that address transportation issues.

4.6 DISCUSSION

The previous sections have highlighted researchers' interest in addressing traffic management problems to minimize road congestion. To measure the performance of suggested solutions, various performance measures have been considered, as listed in Table 4.2. For route guidance and traffic light control methods, these objectives include reduced traffic jams, crowd management, throughput maximization, average waiting time minimization, and average fuel consumption minimization. Routing mechanisms aim to offer better alternative routes quickly to drivers, balancing the benefits of both individuals and groups. This can lead to a nearly balanced

TABLE 4.2 Analysis of Smart Transport for Traffic Management

Objective					Technology				Data Collection			Network Type		Year	Ref
Reduced Traffic Jams	Crowd Management	Throughput maximization	Avg. waiting time minimization	Average fuel consumption minimization	AI	IOT	Big Data	Cloud Computing	Camera	Location services	Historical maps	Isolated intersection	General intersection		
						■			■				■	2016	[17]
			■			■			■			■		2017	[18]
					■							■		2018	[19]
		■				■			■				■	2019	[20]
					■					■				2019	[21]
■				■		■		■					■	2019	[22]
	■				■						■			2020	[23]
			■			■			■			■		2020	[24]
■													■	2020	[25]
			■										■	2020	[26]

traffic load on the entire road network, resulting in the resolution of congestion problems.

Multiple methods have been employed to address the issue of congestion through the intelligent traffic control module. Among them are the game theory approach, multi-agent system, and AI techniques such as fuzzy logic, neural network, deep learning, reinforcement learning, and deep reinforcement learning. Each approach has its distinct characteristics, irrespective of the congestion problem-solving method adopted. Table 4.2 indicates that researchers usually propose traffic light control solutions for individual intersections. However, in practical solutions, intersection coordination methods must be employed to prevent the problem of congestion from arising in other nearby intersections [16].

As previously discussed, AI-based algorithms are used to process and analyze data generated by various devices installed in vehicles to build smart transportation systems. This study focuses on four sub-systems: smart traffic management, smart safety management, smart public transport, and smart manufacturing and logistics. Tables 4.3 through 4.6 summarize the detailed work done on these sub-systems.

Table 4.3 shows that smart traffic management systems can be created by tracking traffic lights in real time during peak congestion hours and suggesting alternative routes. This technique reduces traffic issues and environmental pollution and helps build smart cities.

Table 4.4 illustrates that AI can predict road accidents by identifying road conditions and alerting drivers of potential issues. This results in a significant reduction in road accidents compared to previous scenarios without AI. Smart vehicles equipped with AI-based technologies can also reduce the damage caused by accidents.

Table 4.5 demonstrates that AI helps drivers, travelers, and pedestrians to better plan their journeys by using AI-enabled devices and getting weather forecasting, traffic congestion, and road route information. Efficient AI-enabled public transport systems also support better planning and decision-making.

Table 4.6 shows that AI-based technology has a significant impact on the automotive industry. Vehicles are equipped with cameras, GPS, and sensors to design smart transport systems. Nowadays, various passenger and commercial vehicles are equipped with AI-based solutions for a better and smarter transport experience.

TABLE 4.3   Smart Traffic Management System

Data source	Issue	AI Role	Benefit	Work
Data gathering from smart phones	Routing	Optimal route suggestions	Time management and time saving	Driver Behavior Monitoring Based on Smartphone Sensor Data and Machine Learning Methods [27,28]
Traffic lights and vehicles	Management of traffic during peak hours	Congestion detection in real time	Understanding lower and traffic patterns	Smart Traffic Light System by Using Artificial Intelligence [29]
Smart transport system	Traffic bottleneck	Air pollution detection	Reduction of Environment pollution	Development of an AI model to measure traffic air pollution from multisensory and weather data [30].
Vehicles data	Upsurge of vehicles on the road	Traffic pattern understanding	Route defining and decision-making	Use of ANN and deep learning to determine traffic congestion for heterogeneous networks [31].

TABLE 4.4   Smart Safety Management System

Data source	Issue	AI Role	Benefit	Work
Sensory data from vehicles	Tiredness of drivers	Activating vehicle in autopilot mode	Prevention of accidents	Use of multiple and integrated sensors for safeguarding of people, vehicles, and goods [32]
Automatic vehicles	Safety problem and less performance	Advanced and precise drive assistance, blind spot determination	Reduction of driver's work	Automatic vehicles lead to safety and less work by drivers to ensure safety [12]
Long distance vehicles	Unknown terrain and long driving hours	Health monitoring of drivers	Accidents prediction	Intelligent in-car health monitoring system for elderly drivers [33]
Real-time data from vehicles	High cost of transmission	Route optimization	Techniques to determine volume of vehicles	A Survey of Autonomous Vehicles: Enabling Communication Technologies and Challenges [34,28]

TABLE 4.5 Smart Public Transport System

Data source	Issue	AI Role	Benefit	Work
AI enables vehicles	Delay in delivery time and variation in place	Improving routes and driving patterns	Increased sales and production	Using predictive intelligence by route optimization [35]
Sensors of roads	Determining Road condition	Automatic sending of notifications and alerts to the concerned officers	Improved road safety and management	Building of intelligent transport infrastructure for passenger and driver safety [9,28]
Real-time data from travelers	Traffic bottleneck	Optimizing routes through ML	Reduced travel time	A Survey of Autonomous Vehicles: Enabling Communication Technologies and Challenges [34]

TABLE 4.6 Smart Manufacturing and Logistics System

Data source	Issue	AI Role	Benefit	Work
Vehicles with AI facility	Increase in logistics cost	Information sharing across different platforms and routes	Reduction in cost across the entire pickup and delivery stations	Real-time information dissemination to appropriate drivers across different routes [36]
Connected automobiles	High maintenance and repair cost	Providing predictive and preventive scheduling for easy maintenance	Improvement in smart vehicles along with their monitoring	Performance of connected vehicles is better than autonomous vehicles with easy user interfaces for drivers [37,28]
Smart transportation	Maintenance requirement	Creation of models based on data from logs and IoT sensors	Rescue from machine failure and risk analysis	Intelligent Vehicle System Problems and Future Impacts for Transport Guidelines [38]
Data from documents and bills	Invoice error, poor verification	Fraud purchase and claim detection	More reliability and high accuracy	A Survey of Autonomous Vehicles: Enabling Communication Technologies and Challenges [34]

4.7 AI AND TRANSPORT CORPORATIONS

As previously mentioned, AI-based technologies are being utilized in four transportation subsystems. Additionally, several competitive and large transportation companies are manufacturing AI-enabled vehicles to better serve the public and government sectors. For example, Transport of London has launched Sopra Steria to provide road traffic information and guide passengers through traffic congestion and bus performance. The Ministry of Transport in Singapore has developed nuTonomy, a self-driving autonomous taxi that greatly impacts public transportation. In India, Telangana Transport has developed a chatbot to answer customer queries related to transport, while the French National Railway Corporation has designed a chatbot to assist travelers with transport-related issues and queries. West Bengal Transport has created a mobile application, Patha Disha, to guide people about seat availability, bus arrival timings, and tracking of vehicles, among other things. Similarly, developed countries such as Canada, Australia, and the UAE (Dubai) make use of AI-enabled devices to create intelligent transportation systems. As AI-based technologies rely on complex data gathered from public experiences and vehicles, ensuring the safety and privacy of individuals is paramount. Thus, legal regulations are necessary to ensure the ethical use of data for the development of technology [39].

4.8 AI ACCOMPLISHMENTS IN TRANSPORT

AI has become an integral part of the current world, and there are many notable applications of AI across the globe. One of the most prominent examples is the US Department of Transportation (DoT), which has established an AI-enabled transportation system that serves the American people and economy by providing safe, efficient, sustainable, and equitable movement of people and goods. Local Motors, a US company, has developed a self-driving vehicle called Olli, which is powered by IBM's Watson IoT technology and takes passengers to requested locations and appropriate sightseeing stops. Uber's Otto is the world's first self-driving truck that can deliver beer within 120 minutes, while Hitachi in Japan has embedded AI in their systems, which significantly reduces power consumption in driving rolling stock. GE Transportation in Germany has also developed sensor-equipped locomotives that can detect objects on the way. It should be noted that the examples mentioned are tested against long routes driving, rather than public transport or passenger vehicles. The United States leads in AI-based transportation due to better resources and road infrastructure.

4.9 CONCLUSION

This chapter highlights the importance of AI in the development of intelligent transport systems. It is evident that AI has the potential to address major challenges in the transportation industry. Machine learning models are used to predict and determine traffic congestion and suggest optimal routes. Developed countries have successfully integrated AI into their transportation systems, while developing countries are investing in IoT and AI infrastructure projects. AI has also played a significant role in road safety, traffic management, public transport, and logistics. However, the processing of large amounts of data in various forms such as text, images, and videos raises concerns about individual safety and privacy. Nonetheless, making appropriate decisions requires complex data processing and machine training. The use of AI applications also raises ethical, social, and legal questions, which calls for government intervention to establish policies and ethical regulations for the safe use of data. The insights presented in this chapter will guide government and transportation industry investments to develop digital infrastructure for the benefit of the public.

REFERENCES

1. I. Poola, "How artificial intelligence is impacting real life every day", *International Journal for Advance Research and Development*, 2 (10), 96–100, 2017.
2. M. Chowdhury, A.W. Sadek, "Advantages and limitations of artificial intelligence", in: *Artificial Intelligence Applications to Critical Transportation Areas*, Transportation Research Board, Washington, 6–10, 2012.
3. D. Gershgorn, "Forget the space race, the AI race is just beginning", *Quartz, World Economic Forum*, 2018. Retrieved October 19, 2019.
4. Business Standard, "5 ways NITI Aayog is using AI to change India", *Business Standard*, March 20, 2018.
5. United States Department of Transportation, September 23, 2019. Retrieved October 20, 2019. www.transportation.gov.in
6. R. Abduljabbar, H. Dia, S. Liyanage, S.A. Bagloee, "Applications of artificial intelligence in transport: An overview", *Sustainability* 11 (189), 2019. https://doi.org/10.3390/su11010189
7. B.D. Muynck, B. Johns, O.S. Duran, "Magic quadrant for transportation management systems", *Gartner*, 2019. Retrieved June 7, 2021. www.gartner.com/doc/reprints?id=1-6FPGZY0&ct=190327&st=sb
8. D. Sustekova, M. Knutelska, "How is the artificial intelligence used in applications for traffic management", *Scientific Proceedings of the XXIII International Scientific Technical Conference "Trans & MOTAUTO 15"*, Zilina, 91–94, 2015. Retrieved October 19, 2019.

9. G. Ho, C. Morlet, *Artificial Intelligence in Mass Public Transport*, Centre for Transport Excellence, Land Transport Authority, UITP, Singapore, 2018. Retrieved October 20, 2019.

10. S.G. Ritchie, "A knowledge-based decision support architecture for advanced traffic management", *Transportation Research Part A: General*, 24 (1), 27–37, 1990, https://doi.org/10.1016/0191-2607(90)90068-H.nana

11. J.Z. Hernandez, S. Ossowski, A. Garcia-Serrano, "Multiagent architectures for intelligent traffic management systems", *Transportation Research Part C: Emerging Technologies*, 10 (5–6), 473–506, 2002. https://doi.org/10.1016/S0968-090X(02)00032-3

12. T. Litman, *Autonomous Vehicle Implementation Predictions*, Victoria Transport Policy Institute, Victoria, 2021. Retrieved July 10, 2021.

13. K. Witkowski, "Internet of things, big data, industry 4.0 – innovative solutions in logistics and supply chains management", *Procedia Engineering*, 763–769, 2017. https://doi.org/10.1016/j.proeng.2017.03.197

14. B.H. Li, B.C. Hou, W.T. Yu, "Applications of artificial intelligence in intelligent manufacturing: A review", *Frontiers of Information Technology & Electronic Engineering*, 18 (1), 86–96, 2017. https://doi.org/10.1631/FITEE.1601885

15. Y.Y. Tseng, W.L. Yue, M.A. Taylor, "The role of transportation in logistics chain", *Proceedings of the Eastern Asia Society for Transportation Studies*, 5, 1657–1672, 2005. Adelaide. Retrieved October 21, 2019.

16. A.A. Ouallane, et al., "Overview of road traffic management solutions based on IoT and AI", *Procedia Computer Science*, 198, 518–523, 2022.

17. M.B. Younes, A. Boukerche, "Intelligent traffic light controlling algorithms using vehicular networks", *IEEE Transactions on Vehicular Technology*, 65 (8), 5887–5899, August 2016. https://doi.org/10.1109/TVT.2015.2472367

18. A. Dubey, Akshdeep, S. Rane, "Implementation of an intelligent traffic control system and real time traffic statistics broadcasting", *2017 International conference of Electronics, Communication and Aerospace Technology (ICECA)*, 2, 33–37, April 2017. https://doi.org/10.1109/ICECA.2017.8212827

19. M.B. Younes, A. Boukerche, "An efficient dynamic traffic light scheduling algorithm considering emergency vehicles for intelligent transportation systems", *Wireless Networks*, 24 (7), 2451–2463, October 2018. https://doi.org/10.1007/s11276-017-1482-5

20. A. Frank, Y.S.K.A. Aamri, A. Zayegh, "IoT based smart traffic density control using image processing", *2019 4th MEC International Conference on Big Data and Smart City (ICBDSC)*, 1–4, January 2019. https://doi.org/10.1109/ICBDSC.2019.8645568

21. H.M. Amer, H. Al-Kashoash, M. Hawes, M. Chaqfeh, A. Kemp, L. Mihaylova, "Centralized simulated annealing for alleviating vehicular congestion in smart cities", *Technological Forecasting and Social Change*, 142, 235–248, May 2019. https://doi.org/10.1016/j.techfore.2018.09.013

22. L. Zhao, J. Wang, J. Liu, N. Kato, "Routing for crowd management in smart cities: A deep reinforcement learning perspective", *IEEE Communications Magazine*, 57 (4), 88–93, April 2019. https://doi.org/10.1109/MCOM.2019.1800603

23. M.B. Younes, A. Boukerche, "Intelligent traffic light controlling algorithms using vehicular networks", *IEEE Transactions on Vehicular Technology*, 65 (8), 5887–5899, August 2016. https://doi.org/10.1109/TVT.2015.2472367

24. D.S. Praveen, D.P. Raj, "Smart traffic management system in metropolitan cities", *Journal of Ambient Intelligence and Humanized Computing*, August 2020. https://doi.org/10.1007/s12652-020-02453-6

25. W.C. Tchuitcheu, C. Bobda, M.J.H. Pantho, "Internet of smart-cameras for traffic lights optimization in smart cities", *Internet of Things*, 11, 100207, September 2020. https://doi.org/10.1016/j.iot.2020.100207

26. M. Eom, B.-I. Kim, "The traffic signal control problem for intersections: A review", *European Transport Research Review*, 12 (1), 50, September 2020. https://doi.org/10.1186/s12544-020-00440-8

27. F. Lindow, C. Kaiser, A. Kashevnik, A. Stocker, "AI-based driving data analysis for behavior recognition in vehicle cabin", *Proceedings of the 2020 27th Conference of Open Innovations Association (FRUCT)*, 116–125, 2020. https://doi.org/10.23919/FRUCT49677.2020.9211020

28. L.S. Iyer, "AI enabled applications towards intelligent transportation", *Transport Engineering*, 5, 100083, 2021.

29. S. Zia, M. Naseem, I. Mala, J.A. Mughal, "Smart traffic light system by using artificial intelligence", *Sindh University Research Journal*, 50, 639–646, 2018. Retrieved July 9, 2021.

30. H.B. Ly, L.M. Le, L.V. Phi, V.H. Phan, V.Q. Tran, B.T. Pham, T.T. Le, S. Derrible, "Development of an AI model to measure traffic air pollution from multisensor and weather data", *Sensors*, 19 (22), 4941, 2019. https://doi.org/10.3390/s19224941

31. O.I. Olayode, L.K. Tartibu, M.O. Okwu, "Application of artificial intelligence in traffic control system of non-autonomous vehicles at signalized road intersection", *Procedia CIRP*, 91, 194–200, 2020. https://doi.org/10.1016/j.procir.2020.02.167. ISSN 2212-8271

32. D.J. Yeong, G. Velasco-Hernandez, J. Barry, J. Walsh, "Sensor and sensor fusion technology in autonomous vehicles: A review", *Sensors*, 21, 2140, 2021. https://doi.org/10.3390/s21062140

33. S.J. Park, S. Hong, D. Kim, I. Hussain, Y. Seo, "Intelligent in-car health monitoring system for elderly drivers in connected car: Volume VI: Transport ergonomics and human factors (TEHF), aerospace human factors and ergonomics", *International Ergon Associates*, 20, 40–44, 2019. Daejeon. Retrieved July 10, 2021.

34. M.N. Ahangar, Q.Z. Ahmed, F.A. Khan, M. Hafeez, "A survey of autonomous vehicles: Enabling communication technologies and challenges", *Sensors*, 21, 706, 2021. https://doi.org/10.3390/s21030706

35. P. Jucha, "Use of artificial intelligence in last mile delivery", *Proceedings of the SHS Web of Conferences, Zilina, Globalization and Its Socio-Economic Consequences 2020*, 1–9, 2021. Retrieved July 9, 2021.

36. C. Cheung, Y.L. Chan, K.S. Kwok, W.B. Lee, W.M. Wang, "A knowledge-based service automation system for service logistics", *Journal of Manufacturing Technology Management*, 17 (6), 750–771, 2006. https://doi.org/10.1108/17410380610678783

37. C.Y. David Yang, D.L. Fisher, "Safety impacts and benefits of connected and automated vehicles: How real are they?" *Journal of Intelligent Transportation Systems*, 25 (2), 135–138, 2021. https://doi.org/10.1080/15472450.2021.1872143

38. H. Anandakumar, R. Arulmurugan, A. Roshini, "Intelligent vehicle system problems and future impacts for transport guidelines", *Proceedings of the 2019 International Conference on Smart Systems and Inventive Technology (ICSSIT)*, 1–5, 2019. https://doi.org/10.1109/ICSSIT46314.2019.8987831

39. L.S. Iyer, "AI enabled applications towards intelligent transportation", *Transportation Engineering*, 5, 10083, 2021.

CHAPTER 5

Smart Education Using Explainable Artificial Intelligence (XAI)

Pooja Sharma

5.1 INTRODUCTION

Ever since 1950, when the concept of "thinking machines" was introduced by Alan Turing, consequently, artificial intelligence (AI) research has made significant progress in various fields, leading to an abundance of literature [1–6]. In the realm of education, emerging technologies have revolutionized teaching and learning methods. The US education sector's AI market is predicted to experience a 48% growth rate from 2018 to 2022, as reported by BusinessWire.com. With the flourishing advancements in AI technology, its use in education has increased, presenting opportunities for online face-to-face teaching learning mechanisms, one-to-one learning, self-motivated learning, continuous assessments, and expressive communications in online, mobile, or blended learning environments. In response to the teacher shortage in the United States, some scholars [7] have even proposed replacing certain teaching roles with AI-powered robots.

AI in education (AIED) systems gather information about learners while they use the system, which can be fetched during online teaching interactions with learners and various activities of examinations and assessments. Recently, some systems have started collecting data from other sources such as cameras, microphones, and wearable devices, expanding the scope of data collection beyond keyboard/mouse/screen actions. Learners may or may not be fully aware of the type of data being collected [8].

DOI: 10.1201/9781003409502-5

In Europe, the General Data Protection Regulation has set a global example by codifying societal values regarding data governance and use. This reflects the growing concern that individuals should have control over how technology uses their data [9]. As a result, learners should have the ability to understand how AI works, its potential effects on them, and whether it is trustworthy. Therefore, AIED systems are an important area where Explainable AI (XAI) is needed [10,11].

5.2 EXPLAINABLE AI

The use of XAI is an emerging method for increasing trust in AI systems. XAI encourages the use of procedures that empower human users to comprehend, possess, and appropriately manage the evolving cohort of artificially intelligent associates. Earlier, the main emphasis of XAI was on the algorithm-centric approach [12]. Machine learning models can be classified based on their interpretability level, which refers to the extent to which humans can comprehend the reasoning behind a decision or replicate the model's behavior. Machine learning models with more explicability is necessary because improved resilience to adversarial perturbations and increased ability to detect and rectify different types of biases are among the benefits of this approach, and they enhance the utilization of relevant variables and accurate causation in model inference [13].

Certain types of models, namely general additive models, rule-based models, and decision trees, are considered to be explainable by design due to their easy and simple structure [14]. However, certain models, namely support vector machines, deep neural networks, and tree ensembles, exhibit complex structures that are not directly explainable. In order to make these models interpretable to humans, post-hoc explainability techniques have been extensively developed in the XAI research community, which is developed to focus on how a particular model makes predictions without modifying the actual structure of the model [15].

If we see into the deep insight of the current trend in XAI, it is observed that XAI is moving towards socially situated XAI, where decisions are made based on a socio-organization perspective. This shift is supported by the social sciences, where explanations using AI are not viewed as a product but as a complete process in which social interactions are involved and knowledge is disseminated and transferred from one person to another, or in other words from explainer to explained [16]. As per the target stakeholder and requirement, human-centered approaches are in demand to produce explanations.

XAI required higher sensitivity for the dissemination of knowledge in education and learning standpoint, where self-motivated learners are the potential addressees. Human–computer interaction practices show that XAI methodologies stand apart as considered in improving user's or learner's trust, sensemaking, and decision-making [17]. Explanation in XAI can serve various objectives such as trust building, improved communication, designing, aesthetics, and reasoning in education. In [15], a detailed and useful review of various types of explanations is performed from a cognitive science perspective, which highlights that user-centric approaches are more beneficial for producing explanations

5.3 EXPLAINABLE AI AND EDUCATION

In educational settings, there are various types of explanations that are employed depending on the stakeholders involved and their specific objectives. The primary reason for providing explanations is to ensure accountability to students, parents, or the government. Additionally, educators use explanations to provide feedback to students, diagnose areas of weakness in a class, and assist parents in supporting their child's learning.

Feedback is another significant medium to provide explanation and is very effective in attaining educational outcomes. Teachers provide feedback to students regarding their presentation on particular assignments, ideas for improvement, and suggestions for self-checking and improvement [18]. This type of feedback is quite beneficial to provide a platform for learning that can develop real field or particular domain expertise, develop self-motivated skills, and achieve real intelligence. Consequently, this methodology helps learners to build positive motivation, with more confidence and self-worth.

Feedback for teachers, on the other hand, serves in evaluating various teaching approaches, step by step learning guides, and different types of support provided to the student to help understand that subject in depth. Teachers use various materials to aid their reflections, such as student grade distribution, class assignation, student inspections, parent–teacher meetings, and peer or remote evaluation [19]. Feedback for teachers is vital to the scholarship of teaching, which positions teachers as learners who construct instructional knowledge, pedagogical knowledge, and curricular knowledge that enable them to conduct productive teaching [20].

Another critical form of explanation in educational organizations deals with data to represent the institution's complete profile and achievements and working of the institution [21]. This includes data from various categories such as teachers' research, data which enhances the image of an educational institute to a great extent, student-centric data, which include enrolments, academic performance, student–teacher ratio, research grants, teaching and non-teaching profile, infrastructural, laboratory details, and more. Business intelligence plays a key role in managing and obtaining deep insights from the aforementioned data, specifically providing the data to administrators and management of the institution.

The ultimate goal of providing explanations in these various contexts is to improve and support effective decision-making and uplift the judgment capabilities of the people attached with education policies.

5.4 DESIGN MODEL OF XAI-ED

Designing human–AI interactions to provide appropriate information that helps people understand AI is a complex challenge. Merely opening the algorithmic "black box" is not sufficient to comprehend the implications of AI in the broader sociotechnical system. Even highly accurate AI models, such as deep learning algorithms, can be difficult for data scientists to understand, let alone individuals without formal data analysis training [22]. This is especially true for end users of data-intensive educational innovations, like students and teachers. To address this challenge, this subsection presents various design approaches and research areas for creating AIED systems.

Design teams follow the critical process of user experience (UX) to create products and systems that provide end users with meaningful and relevant experiences. It is to be noted that UX is subjective, which contains the perception of utility, convenience, competence, and ease of use of resources [23]. The design attributes contributing to UX are objective, allowing researchers and designers to maximize opportunities for effective experiences with a system. In order to develop an effective XAI, UX design plays a dominant role. Moreover, recent study reveals that by incorporating explanation and reasoning to the AI systems, they effectively support people in effective decision-making, more accountability and transparency into the existing systems.

However, designing an AI-enabled system comes with numerous challenges that are not otherwise present in other frameworks. To meet the requirements of end users, UX designers are required to be joined with AI developers effectively to design superior models to face challenges of evaluation, technical feasibility, and high impact with an appropriate outcome [24]. Unfortunately, while building an XAI system, UX design is not considered to be the major component along with AI-enabled system development. To address this issue, during the development cycle phase, UX designers must be involved to understand the system from the fundamental perspective and their suggestions incorporated at the initial stage itself, which definitely helps improve system explainability to end users. This simply requires literacy and knowledge of AI capabilities, which are to be developed and incorporated along with core design knowledge and practices [25].

User-centric design (UCD) is another paradigm that is cumbersome to integrate with AI-based teaching and learning mechanism and its analytics. The designers of UCD must be able to understand the requirements of students, teachers, and all involved stakeholders within the institution while designing XAI [26]. Even after the employment of proper and accurate inclusion of UCD with AI, it is not guaranteed that the end users will thoroughly understand XAI explanations along with design needs to provide an understandable and logical explanation about the AI algorithm or its output based on educators'/students' needs and competences.

The designers of XAI-ED can explore top-down guidelines for selecting effective XAI techniques that may function better in a specific domain. For example, researchers proposed different classifications to help other researchers by considering specific machine learning algorithms [27]. Educational theory can build the appropriate and essential foundations for the design of AIED, teaching-learning analytics, and the proper interpretation of their results. In order to support explanation and reasoning, conventional educational theories can be clubbed with XAI designs [28].

The subsequent type of design is participatory design. Participatory design along with co-design helps to build a very effective design process for the stakeholders, where both end users and designers are equally involved. Here the role of the end user is very significant rather than just being a passive object that acts only at the end. This is due to their full grip on their own requirements and experiences, or we can say that they are expert in their own domain [29]. In XAI and teaching/learning system design, if equal participation of educational stakeholders is involved instead of just interview-based system requirement specification based phase, then a very effective system can be developed. However, this can bring certain new challenges also, as an untrained stakeholder without the knowledge of AI and data analysis can give input for the building of an XAI-ED. To address this, an understanding between XAI designers and stakeholders can help build a proper system with identifying appropriate types of explanations [30].

5.5 INADEQUACIES AND SOLUTIONS

As discussed above, although XAI-ED has numerous benefits, it is important to be aware of the potential challenges and pitfalls. This section outlines some of the main pitfalls, along with possible resolutions and ethics for evading them. One such pitfall is the unnecessary use of typical models. It is a usual mistake to employ overly complex models in situations where interpretable models could have yielded similar or better performance. For instance, Gervet et al. [31] found that logistic regression provides far better results than deep learning models in datasets of average size or with a large number of communications per student in the context of student modeling and knowledge tracing. Therefore, it is suggested that designers should emphasize on simple and understandable models and gradually increase intricacy while having a look on precision and interpretability.

However, it is also found that in some cases more complex models provide better results than interpretable models. In a similar illustration, in [31] we see that the use of deep learning approaches outperformed more interpretable approaches for large datasets or in cases where detailed temporal evidence was crucial. In such situations, it is suggested to inculcate complex models with simple models to make hybrid systems to ensure appropriate explainability. For instance, Ghosh, Heffernan, and Lan [32] propose a learner model, which is a synthesis of both complex attention-based neural network models with a set of innovative, interpretable model components used for reasoning based and psychometric models for better understanding and interpretation.

Although explanations are quite important to develop an XA-IED, incorrect explanations can lead to damaging consequences for the stakeholders. And the principal causes of incorrect explanations are the selection of poor or incorrect model, noisy data, and under- or over-fitting. Let us consider an instance, for example, overestimation by lenient open learner model in which student mastery levels are demonstrated as an inaccurate sense of confidence and bad performance in regular assessments and exams. Therefore, it is wise to select a model by rigorous exercise and experiments. Apart from that, deep examination of AI tools is a must in order to improve the quality of an XAI-ED system.

Moreover, as the demand for model understandability increases, some designers of AI tools may be motivated to deliver reasonable but inaccurate explanations. To address this problem, it is crucial to inculcate well-designed explanations relying on well-established mathematical functions that accurately describe the system.

One more disadvantage is unfinished explanations. This occurs due to the high complexity of AI models. Due to this, certain developers provide inaccurate and more simplified explanations that do not solve the real purpose of XAI-ED and conceal the intricacies of adopted model, which again directs to adverse outcomes and an inaccurate understanding of the system. This issue can be resolved by providing an entire set of explanations, including all the aspects without concealing any of the AI model's details. On the other hand, if any of the explanation is incomplete, raising it with proper flagging can make cautious the end user or stakeholder. For many complex XAI-ED systems, it's possible that some users may struggle to comprehend complete and accurate explanations. Kulesza et al. [33] suggest using iterative explanations that are concise and easy to understand, allowing users to incrementally build a more accurate mental model of the system. These iterative explanations can either focus on breadth or depth, depending on the user's needs.

However, there is a concern that providing information on how the system makes decisions or recommendations may lead some users to alter their behavior to achieve a more favorable outcome. For instance, students may repeatedly submit assignments with minor changes to bypass plagiarism detection software, rather than practicing truthfulness and non-involvement in academic delinquency. This matter originates from how the system is developed as compared to its AI performance and accountability. To mitigate this problem, support of game theory methodology could be encouraged. Such practices could inspire users to evade malevolent gaming and undermined conduct [34].

5.6 CONCLUSION

AI is being incorporated in almost every field, and among them education is one of the principal domain where it is being used. As AI algorithms and tools are overtaking every aspect of each domain, people's insecurities are reasonable. Explainable AI is one of the key area where various issues are fixed, such as integrity, accountability, interpretability, and being transparent. In education, XAI plays a significant role in addressing problems concerning students' and educators' independence, perception, fair assessment criteria, academic veracity, and integrity.

In this chapter, various design ideas to develop an optimal XAI-ED are discussed, which include UX, that is, design based on user experiences, which help people to take accurate decisions. Other concept design is based on UCD, that is, user-centric design that signifies that the involvement of learner, educator, institution, and stakeholders is of utmost importance while designing an interpretable and explainable AI system. It is found that the feedback mechanism from both learners and educators also plays a significant role while developing an efficient XAI-ED system. Thereafter, certain challenges and pitfalls of XAI are also discussed. Overall, the chapter discusses the potential and advancement of XAI in the education system, which if implemented in an appropriate manner can help build a more effective and suitable education system as per the current scenario.

REFERENCES

1. Andriessen, J., & Sandberg, J. (1999). Where is education heading and how about AI. *International Journal of Artificial Intelligence in Education*, 10(2), 130–150.
2. Burleson, W., & Lewis, A. (2016). Optimists' creed: Brave new cyberlearning, evolving utopias (Circa 2041). *International Journal of Artificial Intelligence in Education*, 26(2), 796–808. https://doi.org/10.1007/s40593-016-0096-x.
3. Kaplan, A., & Haenlein, M. (2019). Siri, Siri, in my hand: Who's the fairest in the land? On the interpretations, illustrations, and implications of artificial intelligence. *Business Horizons*, 62, 15–25. https://doi.org/10.1016/j.bushor.2018.08.004.
4. Legg, S., & Hutter, M. (2007). A collection of definitions of intelligence. *Frontiers in Artificial Intelligence and Applications*, 157, 17–24.
5. du Boulay, B. (2016). Recent meta-reviews and meta – analyses of AIED systems. *International Journal of Artificial Intelligence in Education*, 26, 536–537.
6. Kulik, J. A., & Fletcher, J. (2016). Effectiveness of intelligent tutoring systems: A metaanalytic review. *Review of Educational Research*, 86, 42–78.
7. Edwards, B. I., & Cheok, A. D. (2018). Why not robot teachers: Artificial intelligence for addressing teacher shortage. *Applied Artificial Intelligence*, 32(4), 345–360.
8. Knijnenburg, B. P., Page, X., Wisniewski, P., Lipford, H. R., Proferes, N., & Romano, J. (2022). *Modern socio-technical perspectives on privacy*. Springer.
9. Wang, D., Yang, Q., Abdul, A., & Lim, B. Y. (2019). Designing theory-driven user-centric explainable AI. *Proceedings of the 2019 CHI Conference on Human Factors in Computing Systems*, 1–15.
10. Drachsler, H., & Greller, W. (2016). Privacy and analytics: It's a delicate issue a checklist for trusted learning analytics. *Proceedings of the Sixth International Conference on Learning Analytics & Knowledge*, 89–98.

11. Holmes, W., Porayska-Pomsta, K., Holstein, K., Sutherland, E., Baker, T., Buckingham Shum, S., Santos, O. C., Rodrigo, M. T., Cukurova, M., Bittencourt, I. I., Koedinger, K. R. (2021). Ethics of AI in education: Towards a community-wide framework. *International Journal of Artificial Intelligence in Education*, 1–23.
12. Gunning, D. (2017). Explainable artificial intelligence (XAI). *Defense Advanced Research Projects Agency (DARPA) and Web*, 2.
13. Miller, T. (2019). Explanation in artificial intelligence: Insights from the social sciences. *Artificial Intelligence*, 267, 1–38.
14. Lipton, Z. C. (2018). The mythos of model interpretability: In machine learning, the concept of interpretability is both important and slippery. *ACM Queue*, 16, 31–57.
15. Srinivasan, R., & Chander, A. (2020). Explanation perspectives from the cognitive sciences – a survey. *Proceedings of the Twenty-Ninth International Joint Conference on Artificial Intelligence Organization*, 4812–4818. www.ijcai.org/proceedings/2020/670.
16. Ehsan, U., Liao, Q. V., Muller, M., Riedl, M. O., & Weisz, J. D. (2021). *Expanding explainability: Towards social transparency in AI systems*. Association for Computing Machinery. https://doi.org/10.1145/3411764.3445188.
17. Alqaraawi, A., Schuessler, M., Weiß, P., Costanza, E., & Berthouze, N. (2020). Evaluating saliency map explanations for convolutional neural networks: A user study. *Proceedings of the 25th International Conference on Intelligent User Interfaces*, 275–285.
18. Hattie, J., & Timperley, H. (2007). The power of feedback. *Review of Educational Research*, 77, 81–112.
19. Boyer, E. L. (1990). *Scholarship reconsidered: Priorities of the professoriate*. ERIC; Brusilovsky, P., Somyürek, S., Guerra, J., Hosseini, R., & Zadorozhny, V. (2015). The value of social: Comparing open student modeling and open social student modeling. In *International conference on user modeling, adaptation, and personalization* (pp. 44–55). Springer.
20. Kreber, C. (2005). Reflection on teaching and the scholarship of teaching: Focus on science instructors. *Higher Education*, 50, 323–359.
21. Drake, B. M., & Walz, A. (2018). Evolving business intelligence and data analytics in higher education. *New Directional Institute Research*, 39–52.
22. Liao, Q. V., Gruen, D., & Miller, S. (2020). Questioning the AI: Informing design practices for explainable AI user experiences. *Proceedings of the 2020 CHI Conference on Human Factors in Computing Systems*, 1–15.
23. Hassenzahl, M., & Tractinsky, N. (2006). User experience-a research agenda. *Behaviour & Information Technology*, 25, 91–97.
24. Yang, Q., Scuito, A., Zimmerman, J., Forlizzi, J., & Steinfeld, A. (2018). Investigating how experienced ux designers effectively work with machine learning. *Proceedings of the 2018 Designing Interactive Systems Conference*, 585–596.
25. Khosarvi, H. et al. (2022). Explainable artificial intelligence in education. *Computer and Education: Artificial Intelligence*, 3, 100074.

26. Shibani, A., Knight, S., & Buckingham Shum, S. (2020). Educator perspectives on learning analytics in classroom practice. *Internet and Higher Education.* https://doi.org/10.1016/j.iheduc.2020.100730.00000

27. Arrieta, A. B., Díaz-Rodríguez, N., Del Ser, J., Bennetot, A., Tabik, S., Barbado, A. et al. (2020). Explainable artificial intelligence (XAI): Concepts, taxonomies, opportunities and challenges toward responsible AI. *Information Fusion,* 58, 82–115.

28. Wang, D., Yang, Q., Abdul, A., & Lim, B. Y. (2019). Designing theory-driven user-centric explainable AI. *Proceedings of the 2019 CHI Conference on Human Factors in Computing Systems,* 1–15.

29. Sanders, E. B. N., & Stappers, P. J. (2008). Co-creation and the new landscapes of design. *CoDesign,* 4, 5–18.

30. Liao, Q. V., Pribić, M., Han, J., Miller, S., & Sow, D. (2021). Question-driven design process for explainable AI user experiences. *arXiv Preprint.* https://arxiv.org/abs/2104.03483.

31. Gervet, T., Koedinger, K., Schneider, J., Mitchell, T. et al. (2020). When is deep learning the best approach to knowledge tracing? *JEDM| Journal Educational Data Mining,* 12, 31–54.

32. Ghosh, A., Heffernan, N., & Lan, A. S. (2020). Context-aware attentive knowledge tracing. In *Proceedings of the 26th ACM SIGKDD international conference on knowledge discovery & data mining* (pp. 2330–2339). Association for Computing Machinery. https://doi.org/10.1145/3394486.3403282.

33. Kulesza, T., Burnett, M., Wong, W. K., & Stumpf, S. (2015). Principles of explanatory debugging to personalize interactive machine learning. *Proceedings of the 20th International Conference on Intelligent User Interfaces,* 126–137.

34. Maskin, E. S. (2008). Mechanism design: How to implement social goals. *The American Economic Review,* 98, 567–576.

CHAPTER 6

Smart Stock Prediction Techniques Using AI and ML

Ochin Sharma, Raj Gaurang
Tiwari, and Heena Wadhwa

6.1 INTRODUCTION

Time series data analysis is the research of data that is collected over time, where the ordering of the data is important. It has applications in various fields, including finance, economics, engineering, and healthcare. The usage of deep learning in time series data analysis has gained popularity due to its capacity of automatically trained from historical information and does accurate prediction. Deep learning is a subset of artificial intelligence that involves the use of algorithms to automatically to create predictions or conclusions, learn from the facts. Time series data analysis using machine learning involves training a model on historical data.

For time series analysis, a variety of machine learning techniques are frequently utilised. For example, regression models such as linear regression and decision trees can be used to simulate how the input and output variables relate to one another. Other algorithms, including support vector machines and neural networks, can be used to capture complex patterns in the data. One of the challenges of time series data analysis using machine learning is the non-stationarity of the data [1–2].

Time series data frequently behaves non-stationarily, meaning that their statistical characteristics shift over time. This might make it challenging to create predictive models since models that perform well during

DOI: 10.1201/9781003409502-6

one era might not perform as well during another. Researchers have created a number of methods, including series data breakdown, trend analysis, and differencing, to solve this problem by making the data more stationary.[3–4] Moreover, more complex machine learning approaches have created to simulate the temporal dependencies including extended short-term memory networks and recurrent neural networks in time series data. Overall, time series analysis using machine learning is a powerful tool. With the potential to produce precise forecasts and provide useful information for decision-making in a variety of domains. It is possible to increase the precision and potency of research of time series data as machine learning techniques advance [5].

Another difficulty in utilizing machine learning to analyses time series data is dealing with missing data. In real-world applications, it is common for time series data to have missing values, which can lead to biased or inaccurate predictions if not handled properly. There are several approaches to handle missing data, including imputation, deletion, and interpolation. Imputation consists of predicting missing values using the dataset. Using the currently accessible data, interpolation includes predicting the missing values. and the underlying patterns in the data. Another challenge is dealing with outliers, which are datasets in the time series that differ greatly in comparison to other data points. Outliers can reduce the precision of machine learning algorithms and can be brought about by errors in data gathering or other circumstances. One approach to handling outliers is to remove them from the data, but this can lead to a loss of information. Alternatively, outlier detection algorithms can be used to identify and flag outliers for further investigation [6].

Also, in time series data analysis using machine learning is model selection and hyperparameter tuning. Numerous useful machine learning parameters and methods can be utilized to model time series data, and selecting the most appropriate algorithm and parameter values can be a time-consuming and complex task. Grid search and random search are commonly used techniques for hyperparameter tuning, while cross-validation can be used to evaluate the performance of different models [5,7,8,6].

In summary, time series data analysis using machine learning has its own set of challenges, including non-stationarity, missing data, outliers, and model selection. However, there are various techniques and approaches that can be used to address these challenges and make accurate predictions. As more data becomes available and machine learning algorithms continue to improve, time series data analysis using machine learning is expected to become an increasingly powerful tool in various fields.

The share market is a complicated system that is susceptible to a wide range of influences, such as changes in political situations, natural disasters, and pandemics around the world. Investors, financial experts, and researchers are all very interested in making predictions about stock market movements. The ability to forecast the stock market accurately can lead to profitable investments and prevent potential losses. Methods of stock market prediction like technical analysis, fundamental analysis, and statistical models have been taken well for decades. For the use of machine and deep learning methods, stock market prediction has gained momentum in recent years [9–12]. The emergence of machine and deep learning has revolutionized many industries, and finance is no exception. The abundance of data and the development of powerful algorithms have enabled machine and deep learning methods to make accurate predictions in many domains. Machine learning enables machines to acquire knowledge with almost no interventions.

First, machine and deep learning methods can handle high-dimensional data, such as financial time series, which is difficult for traditional methods. Second, deep learning and machine learning techniques can identify intricate patterns and relationships in data, which traditional methods may not capture. Third, machine and deep learning methods can adapt to changes in the data, making them more robust to changes in the stock market [13,14].

6.1.1 Conventional Techniques for Foretelling About the Stock Market

Traditional methods of stock market prediction have been used for decades and include technical analysis, fundamental analysis, and statistical models. Technical analysis involves the use of charts and other tools to analyze past price movements and identify trends and patterns. Fundamental analysis involves the analysis of financial statements, company news, and other factors that affect the value of a company. While traditional methods have been used for decades, they have several limitations. First, technical analysis is subjective and can lead to conflicting interpretations of the same data. Second, fundamental analysis may not capture all the relevant factors that affect the stock market. Third, statistical models assume a linear relationship between the past and future values, which may not be the case in the stock market, where complex patterns and relationships may exist [15].

6.1.2 Machine Learning Methods for Stock Market Prediction

A majority of forecasting for the stock market has been carried out recently using machine learning methods. Decision tree algorithm, random forests, vector machines, neural networks, and regression are a few of these techniques. Regression models are straightforward models that assume a linear dependence between the input characteristics and the output variable, such as logistic regression and linear regression. Non-parametric models that can address non-linear correlations and interactions between the input data include decision trees and random forests. A potent classification method that really can handle high-dimensional information and non-linear choice limits is the support vector machine [15].

One advantage of machine learning methods is their ability to handle high-dimensional data, such as financial time series, which can contain hundreds or thousands of features. Machine learning methods can also adapt to changes in the data and make predictions in real-time, which is important in the fast-paced world of finance [16].

However, machine learning methods have one major limitation and that is their lack of interpretability, where it may be difficult to understand why the model makes a particular prediction. Interpretability is important in finance, where decisions can have significant consequences [15,16].

6.1.3 Stock Market Prediction and Deep Learning

Deep learning methods for stock market prediction. Deep learning models can handle sequential data and learn temporal patterns in the data. RNNs (shown in Figure 6.1) have a memory component and can process sequences of data, such as time series. CNNs are commonly used for image processing, but can also be used for time-series analysis. A simple RNN can be represented by the following equation as per Elman:

$$h_t = \sigma_h \left(W_h x_t + U_h ht_{t-1} + b_n \right)$$
$$y_t = \sigma_y \left(W_y h_t + b_y \right)$$

Where h_t : $Hidden\,Layer,$

$y_t = Output\,Layer,\ \ W,U,\ \ b\,are\,attributes,matrics,vector\,respectively$

The most frequently employed deep learning algorithms for forecasting the stock market is the long-short-term memory (LSTM) network.

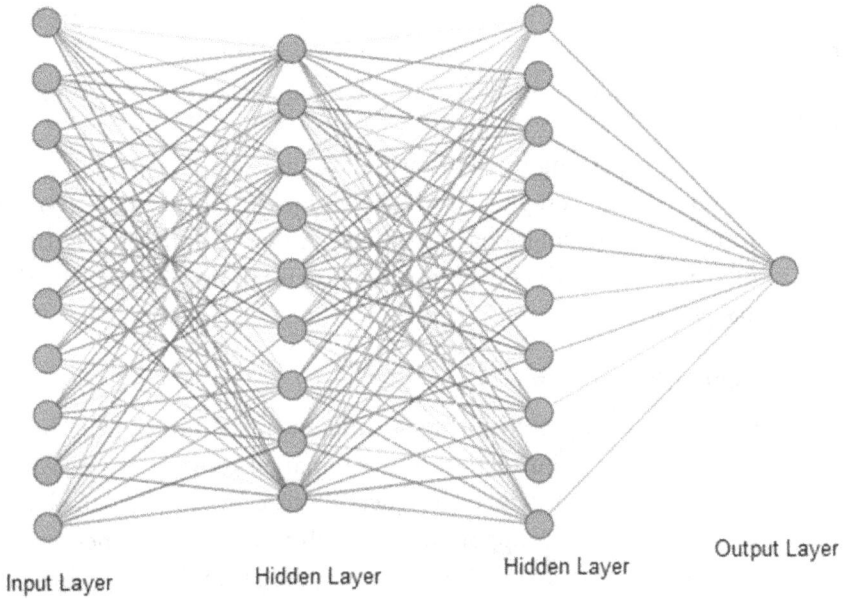

FIGURE 6.1   Deep neural network.

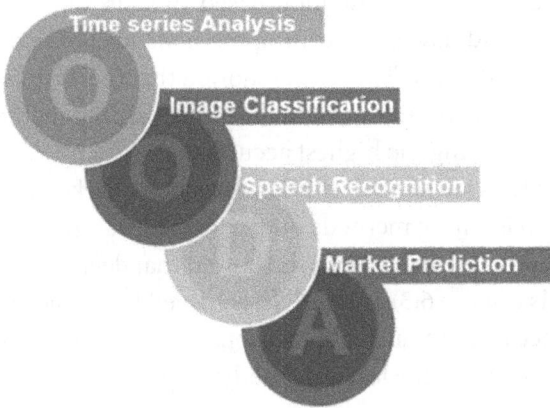

FIGURE 6.2   Applications of deep learning.

In contrast to conventional RNNs, LSTMs do not experience the vanishing gradient issue. The effectiveness of LSTMs for forecasting time series, including stock market prediction, has been demonstrated [17,18].

The applications of deep learning are shown in Figure 6.2. One advantage is their ability to handle sequential data and learn temporal patterns in the data. Another advantage is their ability to learn complex representations of the data, which can lead to better predictions. Deep learning methods can also adapt to changes in the data and make predictions in real-time.

However, deep learning methods also have limitations. One limitation is their computational complexity, which can make them difficult to train and require high computational resources. Another limitation is their lack of interpretability, which can make it difficult to understand why the model makes a particular prediction [17].

6.1.4 Comparison of Performance

In a research [10], the effectiveness of different machine learning techniques for predicting the direction of stock market movement was compared with that of conventional techniques. These techniques included tree structure, random forests, support vector machines, and neural networks. The outcomes demonstrated that machine learning techniques performed better than conventional techniques, with neural networks obtaining the highest levels of accuracy.

Another research compared the performance of various deep learning models, including LSTMs, attention-based models, and hybrid models, with traditional and machine learning methods for predicting the direction of the S&P 500 index. The results showed that the deep learning models outperformed the traditional and machine learning methods, with the hybrid model achieving the highest accuracy.

A meta-analysis of several studies on stock market prediction also found that deep learning methods outperformed traditional and machine learning methods. The meta-analysis showed that deep learning methods, such as LSTMs (Figure 6.3) and CNNs, achieved higher accuracy and better performance metrics, such as precision, recall, and F1-score, compared to traditional and machine learning methods [18].

6.1.5 Applications of Stock Market Prediction

Stock market prediction has several applications in finance and business. One application is portfolio management, where investors can use stock market predictions to make informed investment decisions and optimize their portfolio. Stock market predictions can also be used for risk management, where investors can use them to identify and mitigate risks in their portfolio. In addition, stock market predictions can be used for trading strategies, where investors can use them to make profitable trades and maximize their returns.

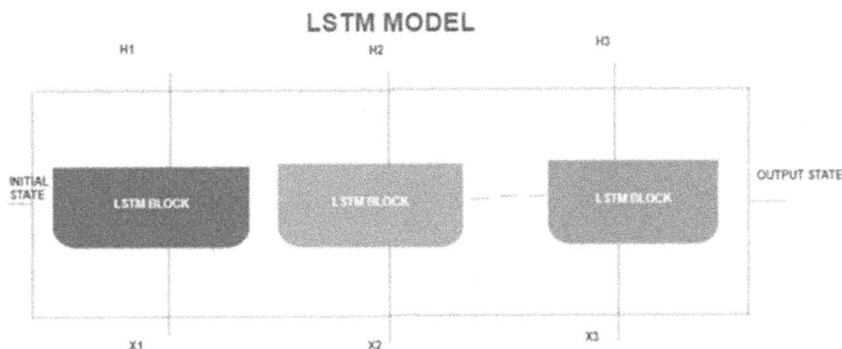

LSTM MODEL

FIGURE 6.3   LSTM model of deep learning.

Stock market predictions can also have broader applications in the economy and society. For example, stock market predictions can be used by policymakers to monitor the health of the economy and make informed decision ns, such as adjusting interest rates or fiscal policies. Stock market predictions can also be used by businesses to make strategic decisions, such as expanding or contracting their operations [17,18].

6.2 LITERATURE REVIEW

The share market is a dynamic, complex system that is impacted by a number of economic, social, and political factors. Predicting stock market prices accurately is an essential task for investors and financial analysts. This literature review provides an summary of research conducted in the area of stock market prediction using ML and DL techniques.

6.2.1 Stock Market Analysis through Recurrent Neural Networks

A form of neural network called neuronal recurrent networks (RNNs) can take into account the temporal dependencies in the data. RNNs have been used in several studies to predict stock prices. A stacked RNN was employed in a research by Li et al. (2019) to forecast the stock closed of a Chinese share market. The outcomes demonstrated that the RNN had a 60.33% accuracy in predicting the closing price.

An RNN was used to forecast the values of a S&P 500 index as well as the Hang Seng index in a different research by Ren et al. (2018). The outcomes demonstrated that the RNN performed better in regards to prediction precision than other ML models, including such SVM and random forests. Similar to this, an RNN was employed to forecast the stock values of the Shanghai Stock Index in a research of Zhang et al. (2018). According to the findings, the RNN performed much better than other ML models as far as of accuracy and stability.

6.2.2 Stock Market Prediction Using Convolutional Neural Networks (CNNs)

Several studies have used CNNs for stock market prediction. A research by Lashkari et al. (2019) used a CNN model to analyze the S&P 500. The results showed that the CNN outperformed other ML algorithms as SVM and random forests, for accuracy prediction.

A research by Wang et al. (2019), a CNN was used to analyze Shanghai Composite Index. The results showed that the CNN performed well over other ML algorithms as RNN and SVM, for both accuracy and stability. Similarly, in a research by Zhang et al. (2019), a CNN was used to predict Hang Seng Index. The results showed that the CNN outperformed other ML models, such as RNN and SVM.

6.2.3 Stock Market Prediction Using Long Short-Term Memory Networks (LSTM)

In a research by Liu et al. (2018), an LSTM network was used to predict closing prices of the S&P 500. The results showed that the LSTM network outperformed other ML models, such as SVM and random forests, in terms of prediction accuracy.

In another work by Zhang et al. (2018), an LSTM network was used to predict the daily closing prices of the Shanghai Composite Index. Comparing results, LSTM network performed well over other ML models, such as RNN and SVM, in terms of both accuracy and stability. Similarly, in a research by Chen et al. (2018), an LSTM network was used to predict the daily closing prices of the Taiwan Stock Exchange Weighted Index. The accuracy results presented that the LSTM network outpaced other ML models like SVM and random forests.

6.2.4 Stock Market Prediction Using Hybrid Models

Hybrid models that combine two or more ML or DL techniques have been used in several studies for stock market prediction. In a research by Kim et al. (2018), a hybrid model that combined an LSTM and a gated recurrent unit was used to analysis the S&P 500. The results showed that the hybrid model outperformed other ML models, such as SVM and random forests, in terms of prediction accuracy.

In a research done by Gao et al. (2019), a CNN and an LSTM network were used to analyze Shanghai Composite Index. The results showed that the hybrid model performed well over other ML models as RNN and SVM, in terms of both accuracy and stability. Similarly, in a research by Yu et al. (2019), a CNN, an LSTM network and a support vector regression (SVR) model was used to analyze Hang Seng Index. The results showed that the hybrid model performed well over other ML models in terms of both accuracy and stability.

6.2.5 Sentiment Analysis and Stock Market Forecast

Sentiment analysis is the process of determining the sentiment of text data. In recent years, several studies have used sentiment analysis to predict stock market prices. A research by Wu et al. (2018), a sentiment analysis was used for Shanghai Composite Index. Results showed, sentiment analysis model performed good over other ML models such as SVM and random forests as per prediction accuracy.

Similarly, research done by Li et al. (2019), a sentiment analysis model was used to analysis S&P 500 index. The results showed that the sentiment analysis model performed well other ML algorithms as SVM and random forests, in terms of prediction accuracy. In another research by Ding et al. (2020), a sentiment analysis model was used to analysis Hang Seng Index. The results showed that the sentiment analysis model outperformed other ML algorithms as SVM and random forests based upon the accuracy.

6.2.6 Reinforcement Learning for Stock Market Prediction

Using actions and incentives, reinforcement learning (RL), a sort of machine learning technique, based on reason to learn from their surroundings. In recent years, several studies have used RL for stock market prediction. In a research by Feng et al. (2019), an RL-based trading agent was developed to predict the daily closing prices of the S&P 500 index. The results showed that the RL-based agent outperformed other trading agents, such as momentum and mean reversion agents, in terms of returns.

Similarly, in a research by Jiang et al. (2019), an RL-based trading agent was developed to analyze everyday closing prices of the Hang Seng Index. The results showed that the RL-based agent outperformed other trading agents, such as the moving average crossover and mean reversion agents, in terms of returns. In another research by Yue et al. (2020), an RL-based trading agent was developed to predict the daily closing prices of the Shanghai Composite Index. The results showed that the RL-based agent outpaced other trading agents, such as the buy-and-hold agent and the mean reversion agent, in terms of returns.

6.3 CHALLENGES IN STOCK MARKET ANALYSIS USING MACHINE LEARNING

Machine learning has shown great potential for stock market analysis, there are several challenges that need to be addressed to make it a more effective tool in the financial industry.

Financial data can be complex and noisy, and there may be missing or incomplete data. In addition, different sources of data may provide different information and may not be compatible with each other. This can make it difficult to obtain a reliable and accurate dataset for training the machine learning models. Another challenge is the non-stationarity of financial data. Financial markets are dynamic and constantly changing, which can make it difficult to develop models that can adapt to new trends and patterns in the data. Models that work well in one market may not work as well in another market, or may perform well during one period of time but not in another period.

Next challenge is the issue of over-fitting. Machine learning models can easily overfit the training data if they are too complex or if the data is too small. As a result, models may perform admirably on training examples but poorly on fresh, untested data. To overcome this challenge, researchers need to use appropriate techniques such as regularization, cross-validation, and ensemble learning to develop models that are more robust to over-fitting. A further challenge is the issue of interpretability. It may be challenging to comprehend how a machine learning model generates its predictions because such models are frequently sophisticated and challenging to interpret.

Finally, there is the challenge of developing models that can account for external factors that may influence financial markets. For example, geopolitical events, news, and changes in government policies can have a significant impact on financial markets, and models that do not take these factors into account may perform poorly.

Machine learning has great potential for stock market analysis. There are in fact many challenges that need to be addressed to make it a more effective tool in the financial industry. Addressing these challenges will require ongoing research and development in areas as data analysis, model interpretability and the development of models that can account for external factors.

6.4 METHODOLOGY

The approach for stock market forecast by machine or deep learning typically involves several steps, which can vary depending on the specific model and dataset used. In general, the following steps are commonly used:

1. Data gathering and pre-processing: The initial step is to collect the relevant data for the stock market prediction task. This can include historical stock prices, news articles, financial reports, and other relevant information.

2. Feature engineering: The next step is to select and engineer the relevant features that will be used as input to the machine learning models. This can involve selecting appropriate time periods, calculating technical indicators, sentiment analysis of news articles, and other data transformations.

3. Model selection and hyperparameter tuning: To utilize the suitable machine-learning models for the task and to tune their hyperparameters to optimize their performance on the training data. This can involve trying different algorithms, architectures, activation functions, learning rates, regularization methods, and other parameters.

4. Training and validation: The fourth step is to sequence the selected models on the training data and to validate their performance on the validation data.

5. Testing and deployment: The final step is to test the best-performing model on the test data, which should be independent of the training and validation data, and to deploy the model for real-time prediction on new data. This can involve monitoring the model's performance over time, retraining the model with new data, and adjusting the hyperparameters or the model architecture as needed.

The methodology (as shown in Figure 6.4) for stock market prediction using machine or deep learning can be complex and challenging, as it requires not only a solid understanding of the underlying algorithms and mathematical models but also domain expertise in finance and economics, as well as practical skills in data manipulation, programming, and statistical analysis. However, by following a rigorous and systematic approach,

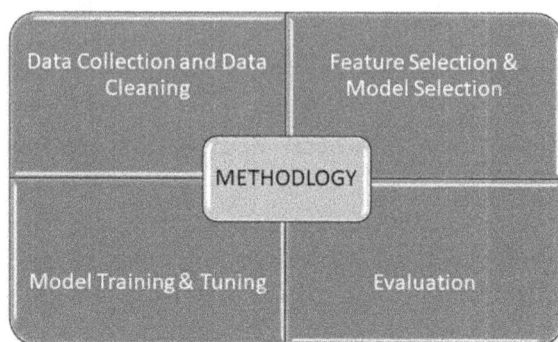

FIGURE 6.4   Overview of methodology of stock prediction.

and by leveraging the latest tools and frameworks in machine learning, researchers and practitioners can achieve promising results.

6.5 CASE RESEARCH AND EXPERIMENTS: PREDICTION OF STOCKS PRICE

Prediction is a reliable method for making assumptions about the future based on evidence already available is prediction. Future decisions will be more advantageous the more accurate the prognosis is. One of the most exciting endeavors has proven to be stock market forecasting. Stock market predictions give investors knowledge about the market, which can help them make wise trading and investment decisions that can result in financial gains.

Processing of data and dispersed communication nodes are crucial in biological systems. Artificial neural networks with inspiration Biological brains and ANNs are different in a number of ways. While the organic brains of the majority of living animals are dynamic and analogue. The employment of several layers by the network is indicated by the term "deep" in deep learning. An inefficient linear feed forward neural structure is replaced by a net with a non-polynomial activation function. Early studies indicate that deep learning is a relatively new technique that uses a limited number of layers with a fixed size to enable practical application and optimization. While concentrating on deep learning, there are numerous considerations to make in order to obtain the optimized results.

6.5.1 Optimization

Optimizers for gradient descent assist in lowering model costs. To cut the cost of the model, the weights are modified using gradient descent.

To identify errors, observations of input and weight combinations are employed. It's also a good idea to descend in steps of the right size, as taking too many steps could skip global minima.

According to deep learning, the main aim of a model is to optimize itself by giving different inputs different weights. Numerous optimizers have been created to employ different weights when improving the model as a result. Because training for large datasets is more time-consuming and resource-intensive whenever a model is changed more frequently with varied weights, choosing an optimizer takes some time. There are many optimizers available, including Adam, Adaguard, Nadam, Adagrad, and SGD. This project's main objective is to use a time series data dataset to test various optimizers in real-world scenarios.

6.5.2 Loss Functions

The difference between the actual and anticipated output is known as the error, and the loss function describes it. The formula that was employed to determine the results is known as the loss function.

$$Actual\ Output - Expected\ Output = F\ (loss)$$

It is calculated to determine the disparity in actual and anticipated production and can be accomplished in a variety of ways. There are consequently many alternative loss functions that can be employed to achieve this. Choosing a good loss function in machine learning is a tough task. Convex functions or loss functions are basically what deep learning utilizes to locate the n-dimensional intervals on a downward convex surface to reduce cost when locating global minima throughout the learning stages. As a result, the model ought to be easily capable of classifying test data. Numerous loss functions are available. For instance, it measures the difference between actual and expected values, or the mean squared error.

- Hinge loss: This is a classification issue The loss function computes the difference between two numbers and makes an effort to maximize that difference. The best classifiers for this situation are support vector machines and their variants. Its convex function is therefore very helpful in machine learning.

6.5.3 Activation Functions

There are a number of activation functions, but they do not all produce findings that are comparable because of variations in statistical design. In general, it has been discovered that when faced with binary classification problems, the best results are obtained when the sigmoid function is utilized as an output activation function. Tanh suffers with a fading gradient issue, so it should only be used occasionally, when dealing with Softmax.

When addressing multi-labeled classification issues with binary classification challenges, it should be avoided. Leaky Relu can be applied when a sizable portion of the input value is zero, because this will lead the Relu function to produce more dead neurons. Making decisions is not possible with dead neurons. As it is cheap, Relu is the most frequently used activation function in networks' hidden layers when calculating costs.

A set of all available input datasets is called an epoch. Each session's weights are examined in order to make a prototype. Weights are adjusted and matched to a subsequent simulation cycle by means of the identical dataset (next epoch). The entire training set must be preserved in the system memory while doing this. Since it is not usually practical to keep the complete dataset in primary memory especially in case of larger datasets, the whole epoch (dataset) is split into batches. The batches are loaded into the primary memory and executed; output is then epochs after the results have been summed together.

If accuracy values are not substantially equal or the simulations are not repeated, increasing the number of experiments with increased epochs is worthless.

When it comes to epoch optimization, there are two important problems. If a system is not able to enhance the models by producing new epochs, this is known as under-fitting. Another is over-fitting, which happens when the system runs out of epochs to simulate in order to make the model better. Increased simulations that don't improve performance lead to over-fitting. Additionally, in this case, the model is trying to optimize weights in an effort to increase accuracy, but this is not feasible given the model's present structural layout.

Based on the stock prediction, these different parameters will be tried with and optimized in this research.

Python platform is being used to simulate the results. Hyperparameters used: Optimizer: Adam, Epochs: 100, Loss Function: MAE, Activation Function: Selu. Apple Inc. dataset is used. The loss value is simulated using data from the previous five years. Single input layer, 3 hidden layers with Sixty neurons each, and one output layer comprise the LSTM deep learning model. Figures 6.5 and 6.6 shows the result of conducted experiments and recorded the loss against activation function and optimizers respectively.

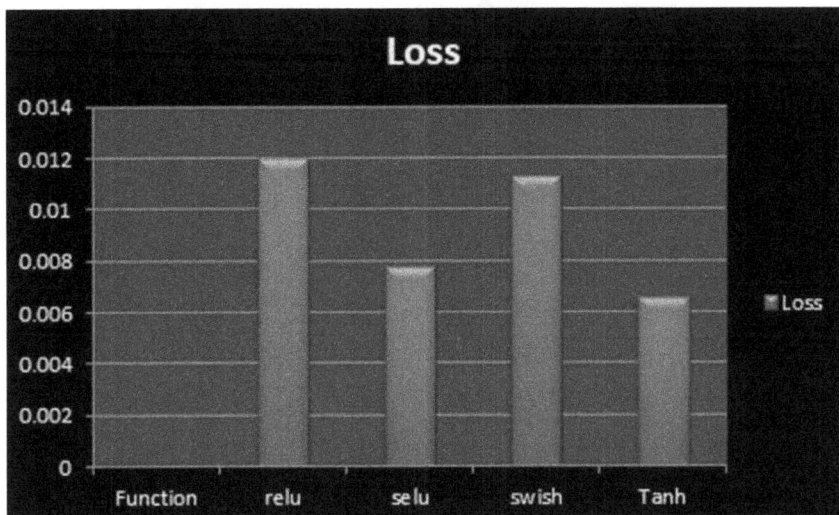

FIGURE 6.5   Loss depicted by various activation functions after experimentation using Python.

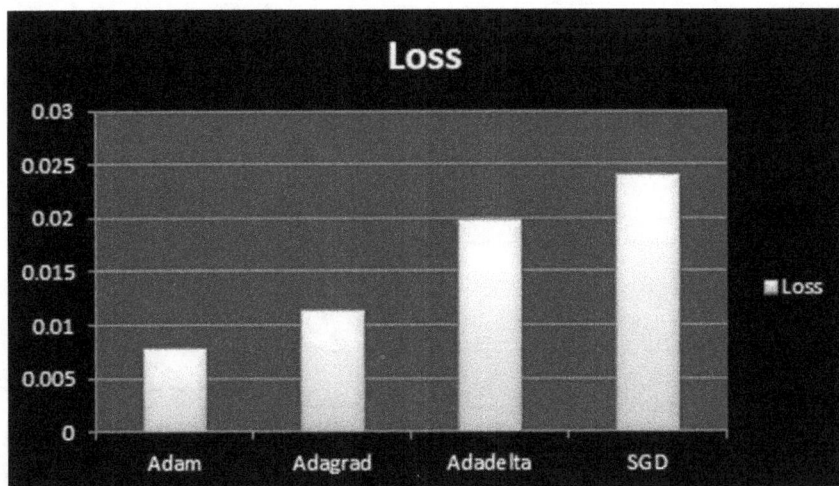

FIGURE 6.6   Loss depicted by various optimizers.

Figure 6.7 shows the historical data of Apple Company. Test size is assumed to be 30% and training size is 70%. The orange line is produced from training data, and the green line is a prediction made from test data.

Figure 6.8 shows the predicted test data (shown in orange) and the predicted training data (green in color).

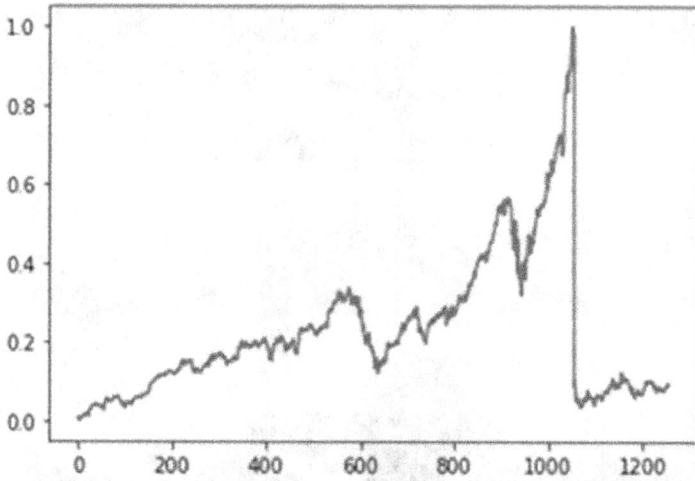

FIGURE 6.7   Historical data from Apple Inc. versus "Close price."

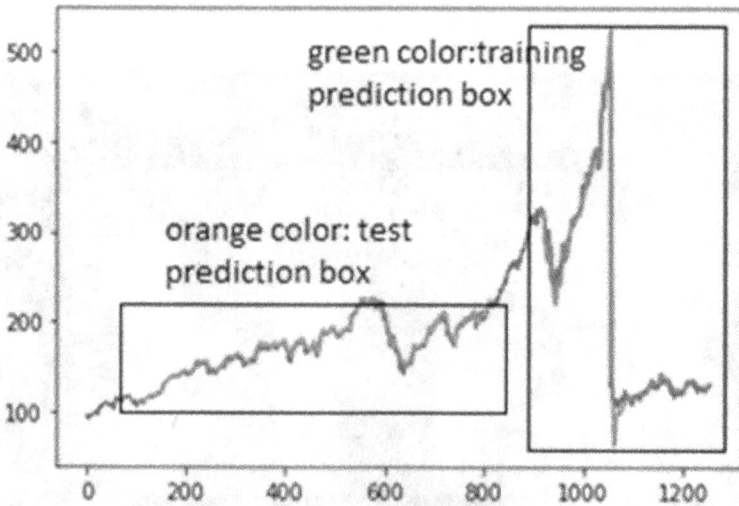

FIGURE 6.8   Apple Inc. stock price prediction.

As a result, it can be deduced from the work done that Adam activation-function and 100 epochs and Selu as an activation function and mean absolute error for instance a loss function to attain optimal accuracy since they serve lower loss values.

6.6 CONCLUSION

The primary objective of this chapter is to show the use of a neural-network for stock price estimate by financial time series information for Apple Inc. stocks. Feed forward neural network has been shown to be effective in predicting time series data. The chapter discusses selection of the loss function, optimizer, and activation function used in the model. MAE is taken as a loss function, adam as an optimizer, and tanh as an activation functional unit, which brought the most promising results. MAE is a commonly used loss function for regression problems and is less sensitive to outliers than other loss functions. The "adam" optimizer is chosen for the model, which is a popular optimization algorithm and has been shown to be effective in many deep learning applications. The optimizer plays an important role in determining the convergence rate and accuracy of the model, and choosing an appropriate optimizer is critical for achieving good performance. For the activation function, the chapter reports that many activation functions are inferior than tanh's activation function in performance. The "tanh" activation function is a recently proposed activation function that deep neural networks have proved to perform well and has several desirable properties, such as self-normalization and preservation of mean and variance. Overall, the chapter provides insights into the choice of loss function, optimizer, and activation function for stock price prediction models. The experiments show that the chosen configurations lead to good performance on the Apple Inc. stock price data and suggest that it is possible to estimate prices more precisely in the future using this model. However, it should be noted that the results may depend on the specific dataset and problem at hand, and further experimentation and validation are needed to confirm the generalizability of the findings.

REFERENCES

1. Huang, D., Shen, D., & Zhao, J. (2020). A comparison of deep learning and traditional machine learning methods in stock market prediction. *Intelligent & Fuzzy Systems*, 38(3), 2633–2643.
2. Jiang, X., Shan, J., & Lu, H. (2018). Stock prediction using LSTM, RNN and CNN sliding-window model. *Journal of Physics*, 1036(1), 012070.
3. Kumar, A., & Lohani, B. (2019). Stock market prediction using machine learning: A systematic literature review. *Intelligent Systems in Accounting, Finance and Management*, 26(3), 129–150.

4. Miao, H., Wan, L., Yang, X., & Cui, L. (2020). Stock prediction based on attention mechanism and gated recurrent unit. *Journal of Computational Science*, 42, 101118.

5. Yao, X., Wang, Y., Huang, J., & Liu, L. (2019). A deep learning framework for financial time series using a mixture model of experts. *Neurocomputing*, 337, 91–102.

6. Jiang, H., Wang, X., & Wu, X. (2019). A transfer learning-based trading agent for stock market prediction. *IEEE Access*, 7, 172425–172436.

7. Li, L., Yang, J., Zhang, H., & Li, Y. (2020). A deep learning model for stock price prediction using geometric Brownian motion and stochastic volatility. *Neural Computing and Applications*, 32(18), 14213–14226.

8. Wang, X., Chen, S., & Chen, Y. (2020). Feature selection method based on the Kruskal-Wallis algorithm for stock market prediction. *International Journal of Advanced Computer Science and Applications*, 11(11), 70–76.

9. Sharma, O., Lamba, V., Pradeep Ghatasala, G. G. S., Mohapatra, S. (2023, June 2). Analysing optimal environment for the text classification in deep learning. *AIP Conference Proceedings*, 2760(1), 020002. https://doi.org/10.1063/5.0150678

10. Wang, J., Li, C., & Li, B. (2020). Deep learning for stock market prediction: A comparative research. *Journal of Ambient-Intelligence and Humanized Computing*, 11(5), 2115–2126.

11. Verma, K., Bhardwaj, S., Arya, R., Islam, U. L., Bhushan, M., Kumar, A., & Samant, P. (2019). Latest tools for data mining and machine learning. *International Journal of Innovative Technology and Exploring Engineering*, 8(9S), 18–23.

12. Feng, X., Chen, Y., & Li, J. (2020). A deep reinforcement learning-based trading agent for stock market prediction using the Hang Seng index. *Journal of Intelligent & Fuzzy Systems*, 38(4), 3835–3843.

13. Wu, Y., & Lai, K. K. (2019). Stock price prediction using support vector regression on daily and up to the minute prices. *Journal of Forecasting*, 38(7), 643–662.

14. Gao, L., & Shi, Y. (2019). Stock price prediction based on a hybrid model of ARIMA and LSTM. *Journal of Physics: Conference Series*, 1168(1), 012044.

15. Pan, S., Wang, R., & Zhang, G. (2020). A transfer learning-based deep neural network for stock market prediction. *Journal of Intelligent & Fuzzy Systems*, 39(1), 1637–1646.

16. Ramesh, T. R., Lilhore, U. K., Poongodi, M., Simaiya, S., Kaur, A., & Hamdi, M. (2022). Predictive analysis of heart diseases machine learning approaches. *Malaysian Journal of Computer Science*, 132–148.

17. Feng, X., Chen, Y., & Li, J. (2020). A deep reinforcement learning-based trading agent for stock market prediction using the Hang Seng Index. *Journal of Intelligent & Fuzzy Systems*, 38(4), 3835–3843.
18. Sharma, O., Lamba, V., Pradeep Ghatasala, G. G. S., & Mohapatra S. (2023). Analysing optimal environment for the text classification in deep learning. *AIP Conference Proceedings*, 2760(1), 020002. https://doi.org/10.1063/5.0150678

CHAPTER 7

Smart Community

Concepts and Applications

Tejinder Kaur and Ojas Sharma

7.1 SMART COMMUNITIES LEVERAGE ADVANCED TECHNOLOGIES

Artificial intelligence (AI) and machine learning (ML) optimize urban services and enhance the quality of life for citizens. They have emerged as key drivers of digital transformation and are playing a vital role in achieving these goals (Abbasi and Abdi, 2018). For instance, AI and ML algorithms can analyze traffic patterns and predict traffic congestion, enabling traffic lights to adjust in real time to improve traffic flow (Abbasi and Abdi, 2018). Smart communities must ensure that citizens' data is collected, stored, and processed securely to prevent any compromise (Abbasi and Abdi, 2018). Additionally, AI and ML algorithms should be designed to detect and mitigate bias and discrimination. Smart communities must also invest in training programs and partnerships with educational institutions to develop the necessary skills and expertise to implement and manage AI and ML systems (Zanella, et. al, 2014).

This chapter explores the concepts and applications of AI and ML in the smart community. It highlights the benefits of using AI and ML to enhance urban services and improve citizens' quality of life (Tang, et. al, 2015). Additionally, the chapter discusses the challenges of implementing AI and ML in smart communities and provides insights into addressing these challenges. By understanding the concepts and applications of AI and ML in the smart community, we can ensure that these technologies are used ethically and responsibly to benefit all citizens (Tang, et. al, 2015).

DOI: 10.1201/9781003409502-7

7.2 SMART COMMUNITIES

Here are some applications of AI and ML in smart communities. Algorithms discover patterns and relationships in the data without any prior knowledge. This type of learning is commonly used for clustering and dimensionality reduction problems (He, et. al, 2016). It analyzes traffic patterns and predicts traffic congestion, allowing traffic lights to adjust in real time to improve traffic flow. AI can also be used to optimize route planning for public transportation, reduce travel time, and provide personalized recommendations to commuters based on their travel history (He, et. al, 2016). AI and ML are transforming the security industry, enabling smart communities to improve public safety and reduce crime. AI and ML are transforming the environmental industry, enabling smart communities to monitor and manage the environment more effectively. ML algorithms can analyze environmental data to identify patterns and predict environmental changes, enabling governments to take proactive measures to prevent environmental disasters (Figure 7.1).

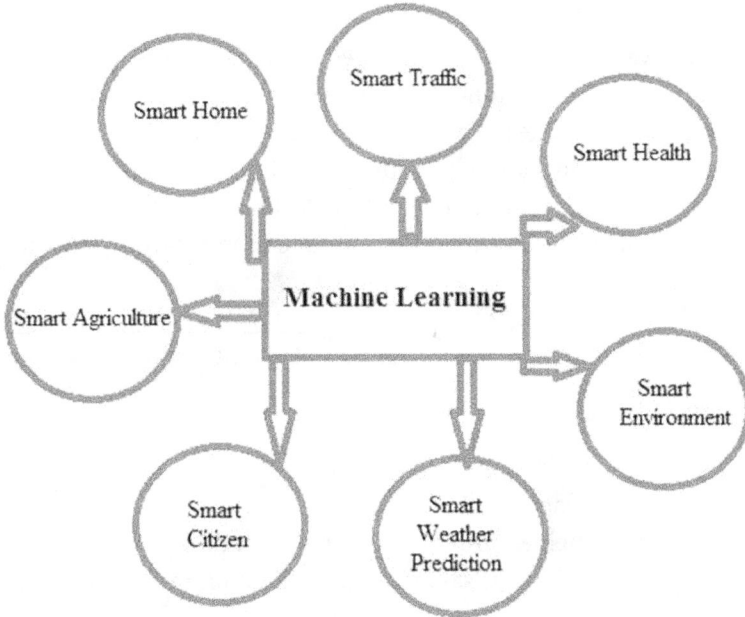

FIGURE 7.1   ML for smart communities.

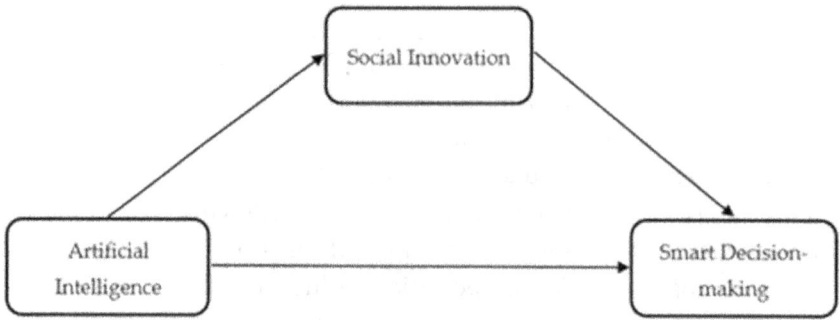

FIGURE 7.2 AI in decision making.

7.3 ARTIFICIAL INTELLIGENCE

AI, in particular, has the ability to automate many aspects of city management, making it easier for administrators to operate and maintain infrastructure while providing residents with better access to services and resources (Li, et. al, 2016).

7.3.1 What Are Smart and Sustainable Communities?

Before we dive into how AI can be used to create smart and sustainable communities, it is important to define what these terms mean. Another area where AI can make a big impact in smart communities is in the area of public safety. This can help to prevent crimes before they occur and improve overall public safety (Li, et. al, 2016). One of the biggest challenges facing sustainable communities is the need to reduce carbon emissions.

7.4 CASE STUDIES

To illustrate the potential of AI to create smart and sustainable communities, let us look at some case studies.

7.4.1 Case Study 1: Singapore

One area where Singapore has leveraged AI is in the area of transportation.

The system has helped to reduce energy consumption by around 20%, resulting in significant cost savings and a reduction in carbon emissions. (Li, et. al, 2016).

7.4.2 Case Study 2: Helsinki

Helsinki, the capital of Finland, is another city that is leveraging AI to create smart and sustainable communities. One area where Helsinki has used AI is in the area of waste management. This has helped to reduce the number of trucks needed for waste collection and has resulted in significant cost savings and a reduction in carbon emissions (Sicari, et. al, 2015). Helsinki has also implemented an AI-powered transportation system that provides real-time information on public transportation options, including bus and train schedules and bike-sharing availability (Sicari, et. al, 2015).

7.4.3 Case Study 3: Portland

Portland, Oregon, is another city that is using AI to create smart and sustainable communities. The city has implemented an AI-powered energy management system that analyzes energy usage patterns consumption. The system has helped to reduce energy consumption by around 25%, resulting in significant cost savings and a reduction in carbon emissions (Sicari, et. al, 2015). Portland has also implemented an AI-powered transportation system that provides real-time information on public transportation options and traffic conditions. This can result in a more efficient and sustainable community that provides better services and a higher quality of life for its residents. (United Nations, et. al, 2014). As we have seen in the case studies above, AI has already been implemented in a variety of ways to create smart and sustainable communities (United Nations, et. al, 2014).

7.5 TRAFFIC MANAGEMENT IN SMART CITIES

Traffic flow optimization systems, such as traffic volume and congestion, help optimize traffic flow.

By using predictive modeling, traffic management systems can identify traffic hotspots and predict when traffic is likely to build up, allowing traffic flow to be dynamically optimized. AI can be used to control traffic signals at intersections, adapting the timing of signals based on real-time traffic data. For example, if traffic is building up on one side of an intersection, the AI algorithm can adjust the timing of the signals to allow more traffic (Li, et. al, 2015).

Route Optimization: By analyzing real-time traffic data, the system can identify the most efficient routes for each vehicle, reducing congestion and improving travel times. AI can be used to predict when road infrastructure, such as bridges and tunnels, will require maintenance. By monitoring real-time data from sensors placed on infrastructure, the system can identify signs of wear and tear and predict when maintenance will be needed (Li, et. al, 2015). By using this information, traffic management systems can take proactive measures to prevent accidents, such as closing lanes or redirecting traffic (Mahmood, et. al, 2016).

7.6 AI FOR SMART URBAN HORTICULTURE

One way AI can be used in urban gardening is through precision irrigation. By using sensors and machine learning algorithms to analyze soil moisture levels, gardeners can water their plants only when necessary, reducing water waste and saving money on water bills. AI can also be used to optimize fertilization, by analyzing soil nutrients and adjusting fertilizer application rates accordingly. Another way AI can be used in urban gardening is through predictive modeling. By analyzing data on weather patterns and plant growth, gardeners can predict when different types of plants will be ready for harvest, allowing them to plan their harvesting and distribution more efficiently. This can help reduce food waste and ensure that fresh produce is available to the community when it is needed (Mahmood, et. al, 2016). By using sensors and machine learning algorithms to analyze pest and disease patterns, gardeners can take proactive measures to prevent outbreaks, such as introducing natural predators or adjusting growing conditions. This can help reduce the need for pesticides and other harmful chemicals, making urban gardening more sustainable and environmentally friendly. It will examine different applications of machine learning, including smart energy management, intelligent transportation systems, and predictive maintenance of infrastructure. Smart energy management is an essential component of creating sustainable communities. This can help to reduce energy consumption and lower energy bills. Another example of machine learning in smart energy management is the use of predictive analytics to optimize energy usage in buildings. Intelligent transportation systems (ITS) are another area where machine learning can be used to create smart and sustainable communities Machine learning algorithms can be used to optimize transportation systems by predicting traffic patterns, optimizing routes, and improving safety. This can help to reduce traffic congestion, lower emissions, and improve safety One example of machine learning in predictive maintenance is the use of sensors to monitor the structural health of bridges. Machine learning has shown great promise in revolutionizing waste management in smart communities. This data can then be used to train algorithms that can predict future waste generation patterns, enabling waste management systems to be optimized for maximum efficiency as more data becomes available and machine learning algorithms become more sophisticated, the potential benefits of this technology are likely to continue to grow (Mahmood, et. al, 2016; Kumar Sachdeva, et. al, 2022a, 2022b).

7.7 CONCLUSION

In conclusion, AI and ML technologies have the potential to greatly enhance the development of smart communities. One of the most significant benefits of AI and ML for smart communities is their ability to make data-driven decisions. Predictive maintenance algorithms can help municipalities identify and fix infrastructure problems before they become major issues. Therefore, it's essential that developers of AI and ML systems ensure that these technologies are designed and deployed in an ethical and responsible manner. In conclusion, AI and ML have the potential to be transformative technologies for smart communities.

REFERENCES

Abbasi, S., and H. Abdi (2018). "A survey on smart cities: Opportunities, applications and challenges," *Journal of Cleaner Production*, vol. 170, pp. 1124–1152.

He, D., S. Zeadally, N. Kumar, and J.-H. Lee (2016). "Anonymous authentication for wireless body area networks with provable security," *IEEE Systems Journal*, vol. 11.

Kumar Sachdeva, R., P. Bathla, P. Rani, V. Kukreja, and R. Ahuja (2022a). "A systematic method for breast cancer classification using RFE feature selection," *2022 2nd International Conference on Advance Computing and Innovative Technologies in Engineering (ICACITE)*, pp. 1673–1676. https://doi.org/10.1109/ICACITE53722.2022.9823464.

Kumar Sachdeva, R., T. Garg, G.S. Khaira, D. Mitrav, and R. Ahuja (2022b) "A systematic method for lung cancer classification," *2022 10th International Conference on Reliability, Infocom Technologies and Optimization (Trends and Future Directions) (ICRITO), Noida, India*, pp. 1–5. https://doi.org/10.1109/ICRITO56286.2022.9964778.

Li, L., R. Lu, K.-K. R. Choo, A. Datta, and J. Shao (2016). "Privacy-preservingoutsourced association rule mining on vertically partitioned databases," *IEEE Transactions on Information Forensics and Security*, vol. 11, no. 8, pp. 1847–1861.

Li, Y., Y. Lin, and S. Geertman (2015). "The development of smart cities in China," *Proceedings of the 14th International Conference on Computers in Urban Planning and Urban Management*, pp. 7–10.

Mahmood, Z., H. Ning, and A. Ghafoor (2016)."Lightweight two-level session key management for end user authentication in internet of things," in *Internet of Things (iThings) and IEEE Green Computing and Communications (GreenCom) and IEEE Cyber, Physical and Social Computing (CPSCom) and IEEE Smart Data (SmartData), 2016 IEEE International Conference on.* IEEE, pp. 323–327.

Sicari, S., A. Rizzardi, L.A. Grieco, and A. Coen-Porisini (2015). "Security, privacy and trust in internet of things: The road ahead," *Computer Networks,* vol. 76, pp. 146–164.

Tang, B., Z. Chen, G. Hefferman, T. Wei, H. He, and Q. Yang (2015). "A hierarchical distributed fog computing architecture for big data analysis in smart cities," in *Proceedings of the ASE Big Data & Social Informatics 2015.* ACM, p. 28.

United Nations (2014). *World Urbanization Prospects: The 2014 revision, highlights. Department of Economic and Social Affairs.* Population Division, United Nations.

Zanella, A., N. Bui, A. Castellani, L. Vangelista, and M. Zorzi (2014). "Internet of things for smart cities," *IEEE Internet of Things Journal,* vol. 1, no. 1, pp. 22–32.

CHAPTER 8

Disseminating Dynamic Traffic Information for Sustainable Mobility and Transport

Jayanthi Ganapathy, Thanushram
Sureshkumar, Ambati Renuka, Vamsi
Krishna, and Ayushmaan Das

8.1 INTRODUCTION

Intelligent Transport Systems (ITS) is a collection of hardware and software components, including roadside infrastructures, CCTV, communication modules, and accessories [1,2]. The components used will considerably vary depending on the services deployed on highways. The recent advancements in ITS are shown in Figure 8.1. ITS services are named as transport telematics in European countries as the services are developed using collective technologies like computation, communication, and information technology [2–5].

European countries were the first to introduce message signs to commuters in the year 2000 for displaying dynamic traffic information. The role of data science was recently established in the transport industry [6–9]. The ITS deployable services are listed in Table 8.1. The implementation of computational algorithms using open source software like Python, OpenCV computer vision module, and R programming has devised solutions for the effective operation of highways nationwide [5,7,8–11].

DOI: 10.1201/9781003409502-8

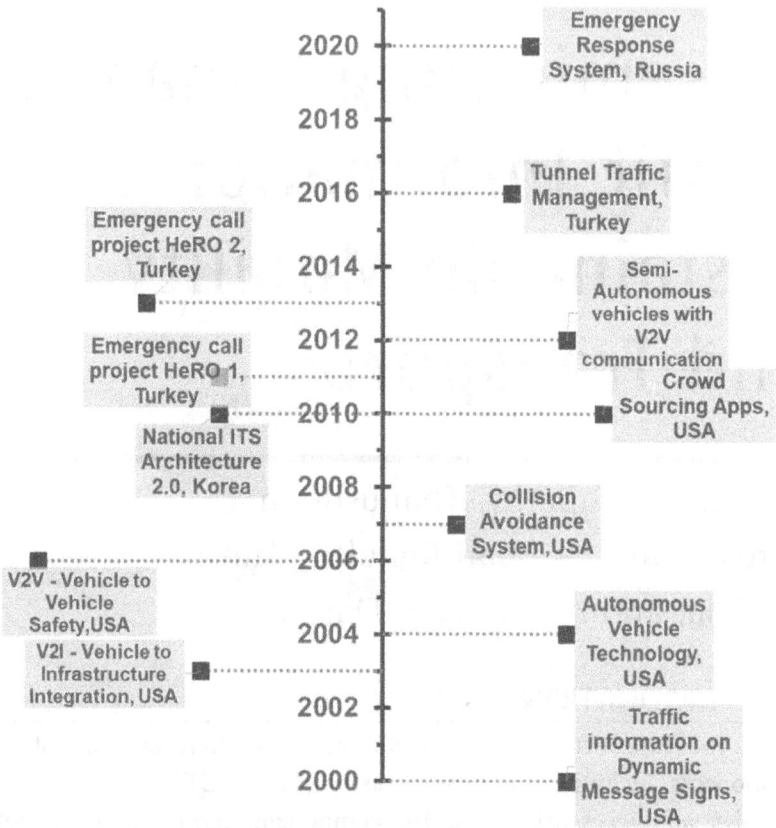

FIGURE 8.1 Worldwide advancements in intelligent transport system.

TABLE 8.1   ITS Deployable Service

Deployable Services	Type of Operations	Role of Data Science
Traveler Information System	Traffic operations	• Artificial Intelligence
Vehicle Control and Safety System	Incident Detection	• Machine Learning and IoT
		• Computer Vision
Traffic Management Systems	Planning and Highway operations	• Data analytics
		• Algorithms
Commercial Vehicle Operations	Multimodal route planning	
Public Transportation System	Metropolitan Transits	

The following section presents the achievements in ITS deployable services in both Indian cities and states of other countries worldwide.

The specific achievements made are that the advent of data communication technologies led the United Nations to launch wireless infrastructure in public transportation through the Vehicle to Infrastructure (V2I) technology, which was introduced in the year 2003 [1,12–14], and the autonomous vehicle technology, which was deployed in the year 2004, followed by safe driving through Vehicle to Vehicle (V2V) communication services in the year 2006 [15–17]. In addition, a system for collision avoidance was deployed in the year 2007. Software development using open-source technology emerged and was introduced to monitor traffic flow on freeways as initiated by the transport department in Minnesota, US. In the year 2010, Koreans introduced their ITS architecture 2.0 [18]. Nevertheless, crowdsourcing apps were developed. The growth of ITS deployment services was tremendous, as seen in the time line beyond 2010 [19–22]. Emergency call and rescue was initiated in Turkey by installing Harmonized e-Call European Pilot (HeERO) versions 1 and 2, followed by tunnel traffic management, which was deployed in the year 2016 [20,21]. Incident management through ERA-GLONASS was deployed by Russians in the year 2017 [23–25]. Smart cities in India emerged from the initiatives of government bodies like the National Highway Authority of India (NHAI). The achievements made by the Indian government are

- A system for freight management using location-based services
- Vehicle tracking through GPS navigation
- Detection of anomalies at stop signals and violations of speed
- Travel information service for public passengers
- Electronic ticket vending
- Public surveillance through cameras
- Emergency response management for reducing causalities and rescue from accidents, mass rapid transits, metro rail transport
- Electronic toll collection through a prepaid tag named FASTag that was deployed in 523 toll plazas all over India by NHAI

Thus, the sole objective of ITS is to ensure safe and secure travel with the help of modern tools and techniques that evolved using information and communication technologies. The worldwide advancements are illustrated with milestones in Figure 8.1, while the advancements in Indian cities are shown in Figure 8.2. The recent advancements in Indian cities after the year 2021 are the development of the Eastern Peripheral Expressway, which was installed on 23 December 2021 [18]. Recent developments on 11 April 2022 are the Onboard Driver Assistance and Warning System (ODAWS), Bus Signal Priority System, and Common SMart iot Connectiv (CoSMiC) software [20]. The Chennai Corporation has proposed a traffic information management system on 17 June 2022.

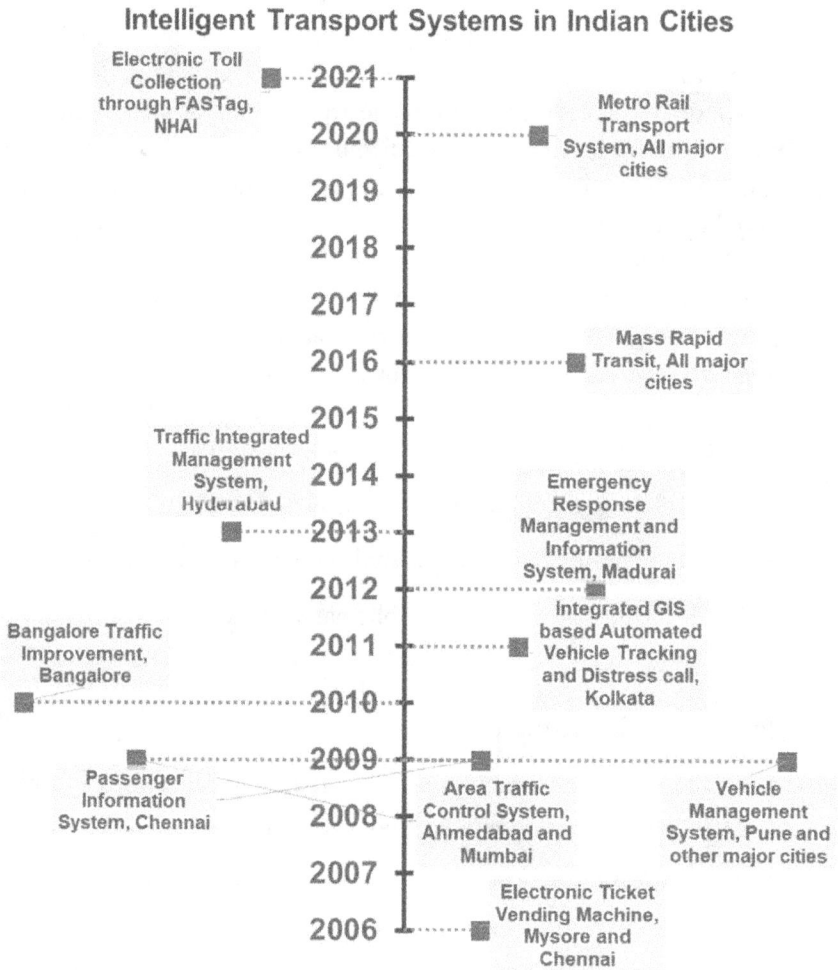

FIGURE 8.2 Advancements in India.

The rest of the sections in this chapter are organized as follows.

- Section 8.2 elaborates the roles of data mining and information technology in transport analytics

- Section 8.3 presents the integral steps used in the development of the application framework for the dissemination of dynamic traffic information

- Section 8.4 illustrates the data collection process and study area used for the evaluation of the application

- Section 8.5 defines the metrics and measures

- Section 8.6 enumerates a detailed case study on the application of data science in ITS

8.2 DATA MINING AND INFORMATION TECHNOLOGY IN TRANSPORT ANALYTICS

The sources of real-time traffic data acquisition are shown in Figure 8.3. Dual loop detectors are used to record the vehicle count [1]. Automated vehicle detection emerged, and later, remote servers and data acquisition networks were deployed to acquire data in different operational settings. The Internet of Things (IoT) has revolved around data acquisition methods and eased the storage and retrieval of real-time traffic flow rates [5].

Temporal and spatial attributes are integral parts of a highway transport network. Modeling the temporal and spatial features of a real-time traffic network would devise a solution for knowledge discovery in travel decisions [1,3,5,7]. Hence, the key components of a traffic information system are as follows:

1. Spatial component: Traffic flow rate at the upstream and downstream

2. Temporal component: Traffic flow rate at time instances

Traffic data acquisition methods have evolved over time, as seen in Figure 8.6. The method used in acquiring traffic counts was initiated with dual loops [1–3,6]. Later, sensor networks were introduced, as reported in recent research outcomes [4,5,13,14,17–21]. However, the advent of technology has also led to outliers in traffic data acquisition. Hence, roadside infrastructures have spurious traffic counts, as discussed in [20]. This can

FIGURE 8.3   Source of traffic data: data acquisition systems and methods.

be handled by spatial-temporal pattern mining using automated vehicle detection systems, as reported in [18–22]. In addition, the advent of data communication technologies has helped devise solutions for streaming the traffic count using big data frameworks [10–15,22–25].

Temporal attributes [1,3,5,7] in a spatial location of the physical transport network represent the traffic instances. Traffic instances are traffic flow rates at time instances, as shown in Figure 8.5. The spatial structures at temporal horizons demonstrate the trend, regularity, seasonality, variability, and randomness, which become the stochastic behavior of the physical traffic flow, as shown in Figure 8.4. Eventually, both spatial and temporal attributes can be quantified at different levels of operational setting in real time. Table 8.2 explains various data acquisition methods, highlighting the role of data mining in transport data analytics. The knowledge gaps identified in traffic flow rate predictions are shown in Table 8.3.

8.2.1 The Key Challenges in ITS

The motivational factors behind the need for data science in the investigation of highway traffic operations are as follows.

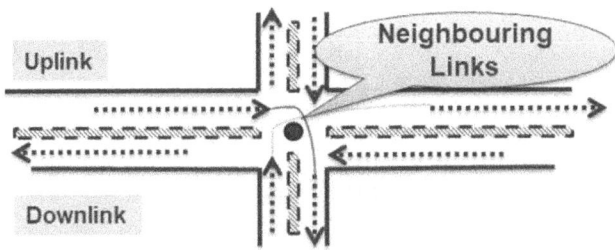

FIGURE 8.4   Spatial component.

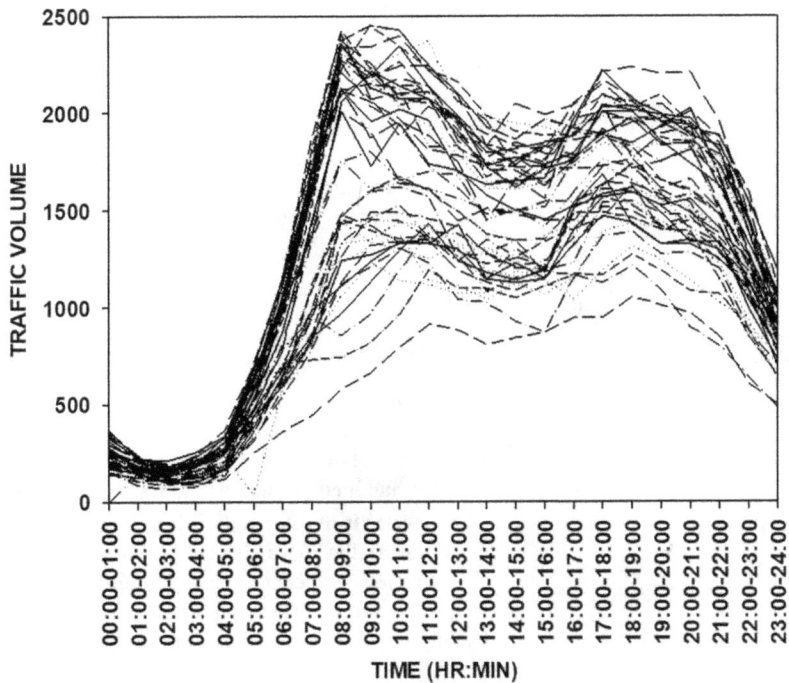

FIGURE 8.5   Temporal component.

1. Content-rich dashboard for profiling daily traffic flow is highly demanded.

2. Accurate prediction with elegant data visualization and dynamic embedding is demanded for effective traffic operations. The time series analysis of univariate and multivariate traffic variables used in travel time predictions are shown in Table 8.5.

TABLE 8.2   Artificial Intelligence and Machine Learning in Transport Analytics

Data Acquisition Methods	Data Mining	# Ref
Software interfacing using hard wired in paved detectors	Travel delay due to queuing of vehicles is investigated.	[1]
Dual Loop detectors	Investigated relative velocity Investigated seasonal patterns	[1–3, 6]
	Non-linear dynamics of traffic flow patterns	[7–12, 16–19]
PeMS Sensor Network	Real-time traffic datasets on a large searchable database	[4, 5, 13, 14, 17–21]
Road side Infrastructure: CCTV	Outlier Detection	[20]
Automated Vehicle Detection	Spatial – temporal patterns	[18–22]
Data and Communication Technologies	Analysis of Traffic drifts using Big data stream, IoT sensors Topological structure for capturing spatial dependence	[10–15, 22–25]

TABLE 8.3   Models and Methods

Models and Methods	Knowledge gap
Model interpretability	Data-intensive computations The complex computational structure Lack of model interpretability hinders practical use of models and methods in different traffic operations.
Outlier analysis	The superiority of computational intelligence in real-time traffic forecasting lies in handling extremeness in fluctuating traffic conditions that would be affected due to short term trends, seasonal changes, cyclic variations and randomness in daily traffic profile.
Prediction of traffic volume, travel time, and estimation of vehicle speed	Temporal and spatial horizons Multivariate analysis Speed up technique for optimizing the time varying networks

3. Improved computational algorithms with consideration on space and time complexity is demanded for online services deploying online

4. Crowdsourcing apps are highly dependent on data from network groups. These limitations have motivated the researchers to focus on traffic data managed through centralized data storage and repository rather than dependency on community user groups.

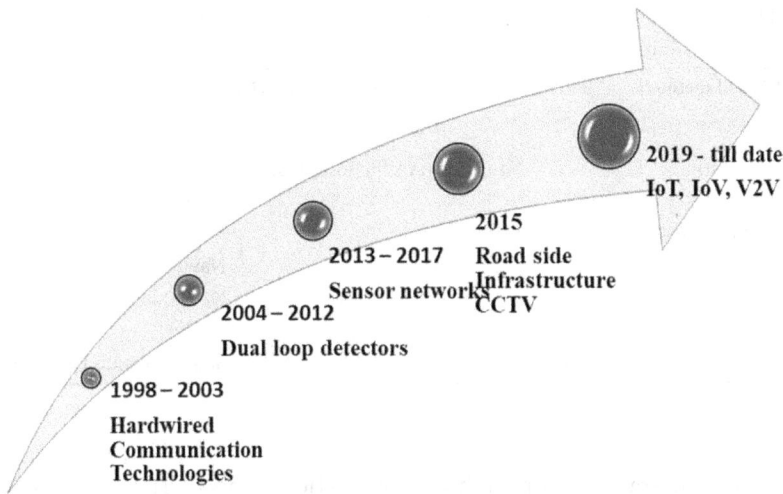

FIGURE 8.6   Traffic data acquisition system.

TABLE 8.4   Traveler Information System: Commercial Proprietary Software in Use by the Public

Mobile App	Year	Ownership	Features	Limitations
Waze	2007	Google Inc.,	Runs on Android OS	Highly dependent on data shared by user on a network
Ola	2010	ANI Technologies Pvt. Ltd. India	Vehicle for Hire Easy affordable trips	Limited to network bandwidth
Uber	2009	Uber Technologies, Inc., USA		Data storage and scalability
Swiggy	2014	Indian Mobile App	India's largest and highest valued online food ordering and delivery Service use for diverse items Food delivery	Limited to network bandwidth
Zomato	2008			Data storage and scalability
Google Maps	2005	Google Incs.,	Real-time traffic condition Speed of vehicle moving in a location.	The number of users from a same location influences the accuracy.

TABLE 8.5 Time Series Analysis in Short-Term Traffic Forecasting with Univariate and Multivariate Variables

Models and methods	Time series analysis	
	Univariate	Multivariate
Deterministic	Short term traffic forecasting	Single step and multi-step forecasting
Non-deterministic	k-nn	k-nn
Machine Learning		Navie Baye's classification
Deep Learning		Neural network, LSTM, RNN
Pattern Mining Techniques		Sequence pattern mining

The modern era is data rich, but computation is poor. The advent of big data analytics has introduced streaming analytics on real-time traffic data. The limitations of commercial proprietary software such as Waze app, Ola and Uber apps, Swiggy and Zomato food delivery apps, and Google Maps, which are used as a travel information service, for navigations are listed in Table 8.4.

8.3 MATERIALS AND METHODS: DISSEMINATION OF DYNAMIC TRAFFIC INFORMATION

TABLE 8.6 Proof by Mathematical Induction

Theorem: The peak hour traffic volume follows the travel time at $t_{k-2} < t_{k-1} < t_k$.

Let $P(k)$ be the basis stated as traffic volume T_v during peak hour following travel time such that $T_v \in t_{k-2} < T_v \in t_{k-1} < T_v \in t_k$.

Traffic volume at peak hour t_k is for the entire duration of given hour k and $|T_v|$ is the number of vehicles flowing through road segment at travel time t.

The value of $|T_v|$ increases with increase in time, thereby causing a delay in travel. Hence, it is true that there is a delay in travel due to traffic volume.

For any given duration of time, it is true that there is a delay in travel according to induction. Hence, $P(k)$ is true. To prove $P(k+1)$ is true.

The amount of traffic at t_{k-1} gets added to vehicles in waiting queue at t_{k-2}.

Hence, the delay in travel time increases the number of vehicles, thereby leading to maximum number of vehicles waiting at t_k the peak hour compared to the previous time instances t_{k-1}, and t_{k-2}. Hence, the stated traffic volume during peak hour follows travel time at $t_{k-2} < t_{k-1} < t_k$.

Models and Methods in Short term Traffic Forecasting

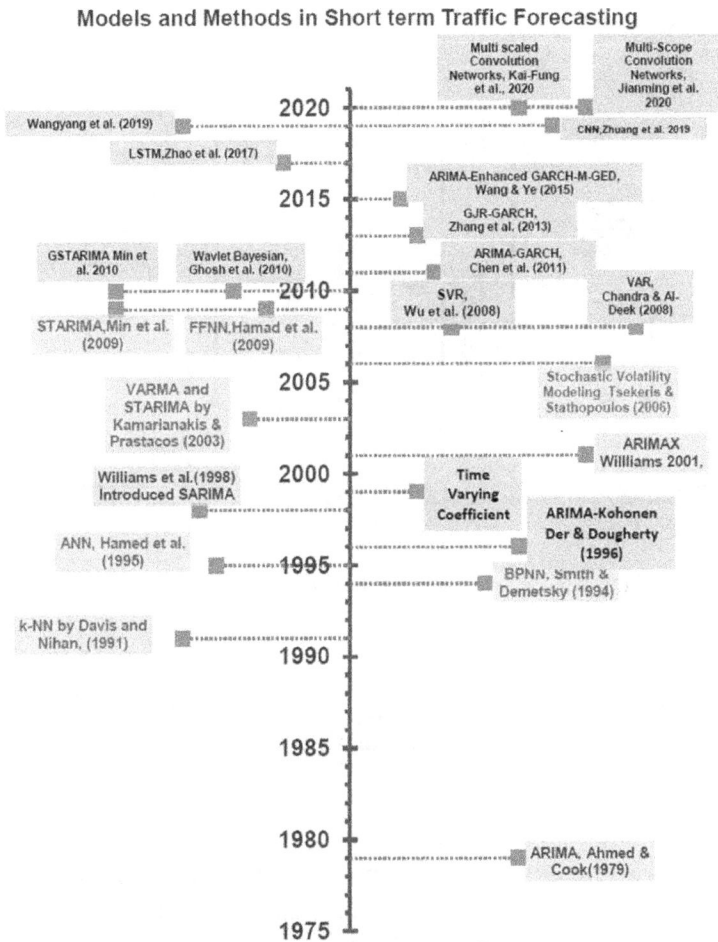

FIGURE 8.7   Models and methods in time series traffic forecasting.

8.4 DATA COLLECTION

The software application framework for profiling the daily traffic involves a sequence of tasks, as illustrated in Figure 8.8. The input, process, and output phases of the online application framework is explained as follows.

Step 1: The real-time raw traffic count is stored in a repository. The repository is connected to all social media services. Hence, users on social media can share the traffic information at any time and in

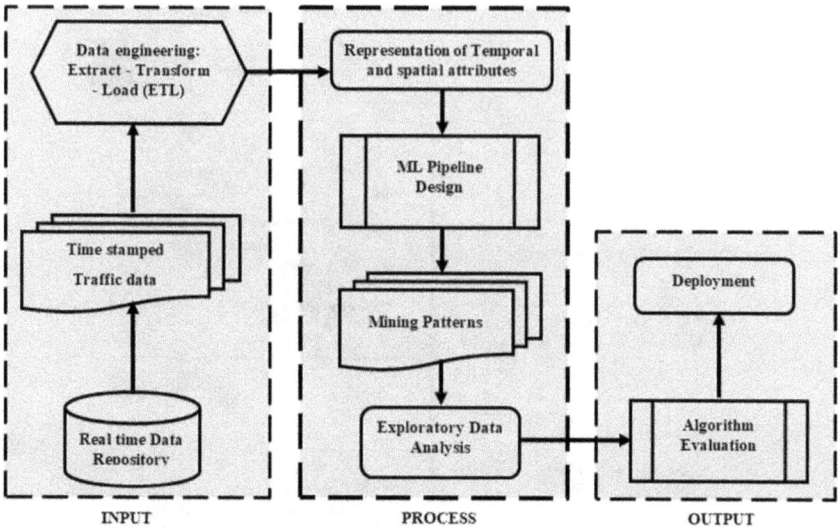

FIGURE 8.8   Flow diagram of application framework.

any location. The apps in a connected media network like Twitter, Instagram, Facebook, and Google are the source of data for the repository.

Step 2: Time-stamped traffic data is extracted from the repository. A data wrangling process is applied on time-stamped traffic counts to convert them to a suitable format for analysis. This process involves a data pipeline design on a cloud storage.

Step 3: The machine learning pipeline is designed to generate patterns for analyzing traffic sequence patterns. The proof of the concept used in generating traffic sequence patterns is shown in Table 8.6.

Step 4: The algorithms used for pattern mining are evaluated and validated before the deployment of the service for public use.

The study area investigated in this study for the experimentation and evaluation of software frameworks is shown in Figure 8.9. These state highways 49A and 49 are in the southern regions of Chennai Metropolitan City, as seen in Figure 8.9. Traffic operation in these state highways are under the governance of IT Expressway LTD (ITEL), itel.tnrdc.com, and the Tamil Nadu Road Development Corporation (TNRDC). The three

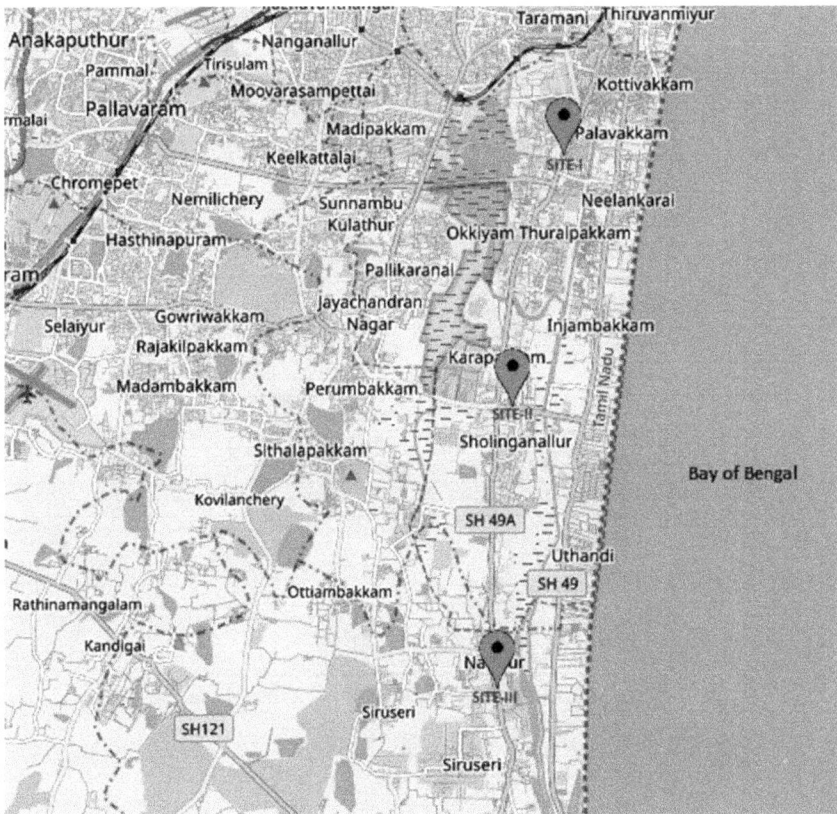

FIGURE 8.9 Experimental site: State Highway 49A and 49.

sites pinned on the image are the toll plazas connected to a centralized server. The real-time traffic counts are captured and stored in a repository on the connected network. The dataset is extracted in the form of portable documents or comma-separated values for further analysis. The various attributes considered for processing the raw traffic count are listed as a data dictionary in Table 8.7. The traffic count on each day is stored in a format containing the date of capture, name of the toll, time and lane id, the count of each type of vehicle, and the passenger car unit. The sample datasheet is shown in Figure 8.10. The vehicle flow in upstream and downstream of a lane in toll plaza is recorded under the fields exit and entry. The time stamp shown in this datasheet is in 24-hour format. Each sheet depicts a 24-hour traffic count, with a count for each vehicle type for each lane. Each hour of the traffic count is an aggregated sum of vehicle flow in

Time	3 WHEELER		CAR		LCV		BUS		TRUCK		MAV		Sub TOTAL		Total Traffic	Total PCUs
	ENTRY	EXIT	ENTRY	EXIT	ENTRY	EXIT	ENTRY	EXIT	ENTRY	EXIT	ENTRY	EXIT	ENTRY	EXIT		
Payment Type :	CASH															
0:00-1:00	1	0	115	30	10	2	0	0	2	0	9	4	137	36	173	222
1:00-2:00	0	0	79	24	0	4	0	0	1	0	5	1	93	29	122	148
2:00-3:00	0	0	41	13	0	3	0	0	4	0	0	0	54	16	70	84
3:00-4:00	0	0	58	34	10	9	0	0	2	1	0	1	70	45	115	134
4:00-5:00	0	0	67	34	13	9	0	0	3	0	2	1	85	44	129	155
5:00-6:00	0	0	104	38	19	4	0	0	7	0	3	0	132	42	174	206
6:00-7:00	1	0	167	50	20	11	24	5	1	0	0	0	213	66	279	355
7:00-8:00	0	0	376	75	35	6	104	3	1	0	0	2	516	86	602	845
8:00-9:00	0	0	562	182	48	6	31	0	0	0	0	0	641	188	829	918
9:00-10:00	2	1	621	207	13	2	3	0	0	0	0	0	639	210	849	863
10:00-11:00	8	0	695	227	14	3	0	1	0	0	0	0	717	231	948	959
11:00-12:00	3	1	635	150	17	2	1	2	0	0	0	0	656	155	811	827
12:00-13:00	3	0	599	132	30	9	0	0	0	0	1	0	633	141	774	797
13:00-14:00	5	1	476	127	25	5	3	0	2	0	0	0	511	133	644	669
14:00-15:00	3	0	478	143	24	5	0	0	0	0	1	0	506	148	654	672
15:00-16:00	3	0	442	116	27	3	1	9	1	0	0	0	474	128	602	639
16:00-17:00	2	0	443	113	37	3	0	27	1	0	0	0	483	143	626	702
17:00-18:00	2	1	474	101	26	5	2	2	0	0	0	0	504	109	613	637
18:00-19:00	0	0	445	100	18	5	1	1	1	0	0	0	465	106	571	589
19:00-20:00	1	0	423	94	12	2	2	0	0	0	1	0	439	96	535	549
20:00-21:00	0	1	499	100	17	3	3	0	0	0	0	0	519	104	623	639
21:00-22:00	1	0	373	121	26	2	1	1	0	0	0	0	401	124	525	543
22:00-23:00	0	0	310	91	14	9	1	0	4	0	3	0	332	100	432	463
23:00-24:00	0	0	198	55	21	3	0	1	4	0	5	0	228	59	287	334

FIGURE 8.10 Traffic dataset sheet.

TABLE 8.7 Data Dictionary

Name of the Attribute	Data Format
Date	dd-mm-yyyy
Name of the toll plaza	Text
Time	Hr:min:sec
Lane id	Number
3-wheeler automobile	Number
Car	Number
LCV	Number
BUS	Number
Truck	Number
MAV	Number
Uplink vehicle count	Number
Downlink vehicle count	Number
Total count	Number
Passenger Car Unit (PCU)	Number

15-minute resolution as per the guidelines of the *Indian Highway Capacity Manual*, 2020 [3].

8.5 METRICS AND MEASURES

The application framework developed for traffic profiling is evaluated using the metrics discussed in this section. The various metrics and measures of traffic predictions used in the evaluation of traffic sequence mining are explained as follows.

8.5.1 Hypothesis Testing

$$\chi^2 = \frac{\Sigma(O_i - E_i)^2}{E_i} \tag{8.1}$$

Here, O_i is the observed traffic volume and E_i is the predicted traffic volume.

8.5.2 Mean Absolute Error (MAE)

$$MAE = \frac{\sum_{i=1}^{N}|Y_i - y_i|}{N} \tag{8.2}$$

Here, Y_i is the experimental measure of traffic volume (the predicted volume), Y_i is the traffic volume in test value (the actual traffic volume), and N is the number of observations.

8.5.3 Mean Relative Error (MRE)

$$MRE(x', x) = \frac{\left(\sum_{i=1}^{N} \frac{|x_i' - x_i|}{x_i}\right)}{N} \tag{8.3}$$

Here, x_i' is the predicted value, x_i is the observed value, and N represents the number of observations.

8.5.4 Sequence Similarity

$$\ell(\alpha, \mathbb{A}) = \frac{|\alpha| - |Cov_{\mathbb{A}}(\alpha)| + 1}{|\alpha|} \tag{8.4}$$

$|Cov_{\mathbb{A}}(\alpha)|$ is the number of subsequences composing an $\mathbb{A}$-optimal covering of α and $|\alpha|$ is the length of sequence S.

8.5.5 Edit Distance

$$lev\_dist_{x,y}(u,v) = \begin{cases} max(u,v), if\ min(u,v) = 0, \\ min\begin{cases} lev_{dist\ x,y}(u-1,u)+1 \\ lev_{dist\ x,y}(u,v-1)+1 \\ lev_{dist\ x,y}(u-1,v-1)+1_{x_u \neq y_v} \end{cases} \end{cases} \tag{8.5}$$

u *and* v are the positions of terminating symbol in sequence x *and* y respectively. If $x\_u \neq y\_v$, the number of count needed to edit the sequence is incremented by 1; otherwise, the number of edits in the sequence is zero.

8.5.6 Cosine Similarity

$$Cos_{similarity}\left(X_s, Y_s\right) = \frac{X_s \cdot Y_s}{|X_s| \times |Y_s|}$$

(8.6)

Here, S_x and S_y are the two non-empty sequences, and X_s and Y_s are their corresponding vectors of equal length.

8.6 CASE STUDY ON INTELLIGENT TRANSPORT SYSTEMS

Computer vision is a subfield of artificial intelligence; it involves a collective framework of image processing algorithms that teaches the computer to capture, analyze, interpret, and extract information from images and videos [19]. The hierarchy of computational intelligence in computer vision-based software system as used in transport analytics is shown in Figure 8.11. Computer vision is a subfield of data science, which is a subfield of artificial intelligence. Hence, the features of open-source software developed using algorithms in computer vision and data science underlie the concepts and principles of artificial intelligence. Computer vision

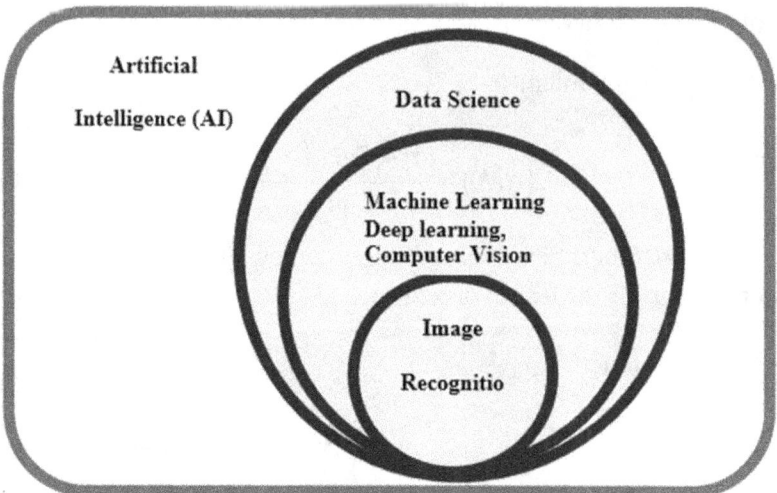

FIGURE 8.11 Hierarchy of computational intelligence in data science.

concepts are implemented using *OpenCV* modules in python, an open-source package for interpretation, storage, selection, and extraction of information from image and video data. The *OpenCV* module is released under GNU public license; it is a freely available software package [20]. This section presents the case study on research methodologies and its application in ITS.

The following section explains the case study on intelligent transport systems.

8.6.1 Case Study 1 – Vehicle Trajectory Mining

Trajectory is a sequence of tasks performed to quantify a physical phenomenon in real time. Vehicle trajectory is a series of vehicles that are moving in a sequence that can be captured from an image or video. Mining such trajectory yields useful information for tracking and navigating vehicles in real time. This in turn helps in daily traffic profiling for public transport and travel decisions. The data collection is performed using the roadside cameras installed on the campus wing of the study area, Sri Ramachandra Institute of Higher Education and Research, Porur, Tamil Nadu – 600 116. The experimentation was performed using the camera(s) numbered 4, 6, and 10. The various tasks involved in vehicle trajectory mining are

1. Object detection

2. Vehicle counting

3. Speed estimation

The YOLO [23–25] model is a deep learning framework used in transport analytics for object detection. The sequence of steps used in object detection using a YOLO deep learning framework is shown in Figure 8.12. The camera module present in the roadside infrastructure captures video of the traffic. This video is segmented frame by frame. The frames are processed by a convolutional neural network for feature extraction [25], the objects in the scene. It draws a bounding box on each detected object to classify it into the types of objects like vehicles (two wheeler, three wheeler), pedestrians walking on the road, and other types of vehicles. The DeepSORT shown in Figure 8.13 is an algorithmic approach to work along with YOLO for the estimation of speed. A line shown in green in Figures 8.14, 8.15, and 8.16 is used as a reference line to calculate the distance traveled by the vehicle in the scene. The output of YOLO is used to estimate the speed.

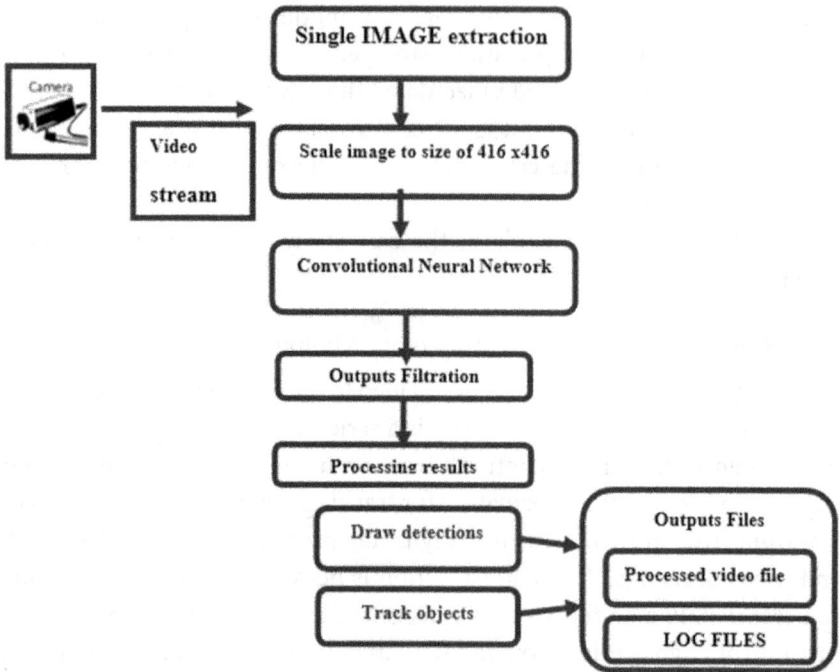

FIGURE 8.12   Object detection using YOLO.

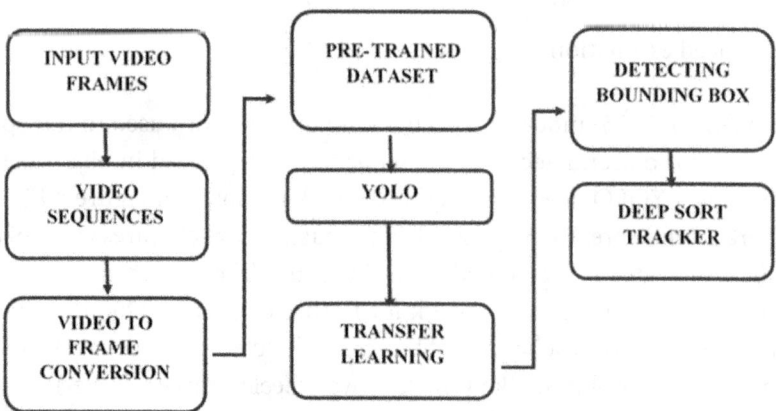

FIGURE 8.13   Deep SORT workflow.

FIGURE 8.14   Camera 4: vehicle count using YOLO model.

FIGURE 8.15   Camera 6: Object detection and vehicle count using YOLO model.

FIGURE 8.16 Camera 10: object detection and vehicle count using YOLO model.

8.6.2 Case Study 2 – Traffic Profiling Studies

The objectives daily traffic profiling are as follows.

1. To capture the dynamics of physical traffic flow by an Extract-Transform-Load (ETL) data pipeline design for the representation of raw traffic count

2. To design a machine learning pipeline that augments the traffic sequence mining framework with vehicle speed based on multi-criteria decision-making support for profiling the highway traffic

3. Design an analytic pipeline to disseminate dynamic traffic information in successive time instances and operate the vehicular traffic with the help of an interactive dashboard

The system is designed to integrate various stages of the data pipeline deployed in this project, as shown in Figure 8.17. The physical traffic flow is captured and transformed to a representational form using the data

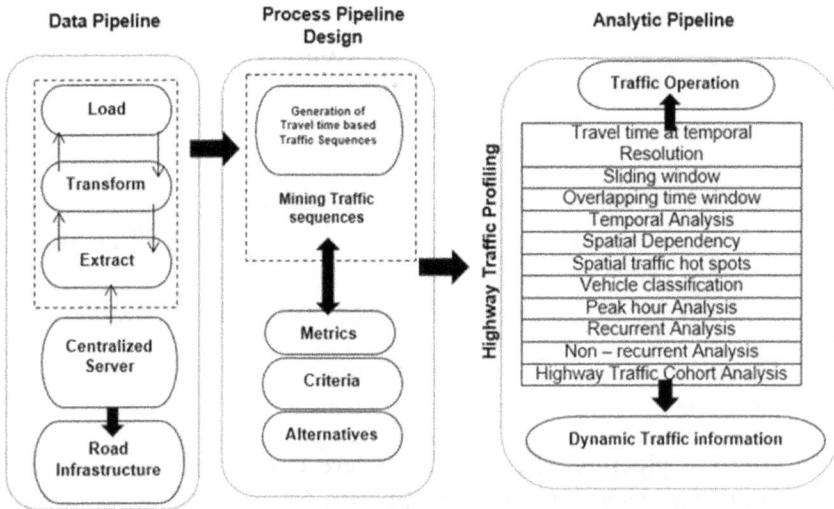

FIGURE 8.17   Highway traffic profiling system: system flow diagram.

pipeline. Travel time-based traffic sequence generation was performed in the process pipeline. These evaluated traffic patterns become the information sources for the formulation of decisions based on alternatives with more than one criteria. The result of pattern generation algorithms is further analyzed by profiling the physical traffic flow at various temporal resolutions, varying the window size ranging from 5 minutes to 30 minutes in the future. Further, highway traffic profiling is influenced by overlapping windows at different temporal resolutions, spatial dependency in different operational settings, and temporal traffic hotspots considering spatial characteristics. In addition, the system shall report the classification of vehicles based on traffic during peak hour, recurrent traffic control, and/or non-recurrent congestion analysis when deployed as mobile and desktop application.

8.6.3 Societal Benefits

Deployable services are one of the principal applications of information technology in ITS. Highway traffic profiling is one such deployable ITS service. Automated software for e-traffic alert systems has its importance as an automated software unit developed for profiling highway traffic. These systems disseminate dynamic traffic information in successive time instances. The software system, when deployed, would assist travelers to commute safely, plan a hassle-free journey on long trips, and plan their travel for work. Following are the essential outcomes, with an emphasis on social relevance.

1. **Data pipeline:** The highway traffic network is a dynamic system. This pipeline returns a representation of the physical traffic flow rate. At times, traffic flow on highways is unprecedented. Thus, the framework would help commuters gain insights into physical traffic flow.

2. **Process pipeline:** The commuters benefit from proactive decisions in planning their travel trips.

3. **Analytic pipeline:** The commuters have access to the analytic dashboard for visual interpretation of spatial and temporal traffic flow rates in traffic operations.

8.7 CONCLUSION

This chapter presents the role of data mining and information technology in transport analytics for deployable ITS services like the dissemination of dynamic traffic information. This leads to proactive congestion management. In this perception the chapter aims at the formulation of integral steps used in the development of an application framework for the dissemination of dynamic traffic information. Further, the data collection process and study area used for evaluation of the application are presented. The metrics used for the assessment of travel time and traffic pattern are explained in detail. Finally, case studies on vehicle trajectory mining and daily traffic profiling are elaborated with illustrative results.

REFERENCES

1. Jayanthi, G & Jothilakshmi, P (2021). Traffic time series forecasting on highways: A contemporary survey of models, methods and techniques. *International Journal of Logistics Systems and Management*, vol. 39, no. 1, pp. 77–110.
2. Zhuang, Y, Ke, R & Wang, Y (2019). Innovative method for traffic data imputation based on convolutional neural network. *IET Intelligent Transport Systems*, vol. 13, no. 4, pp. 605–613.
3. Jayanthi, G & Jothilakshmi, P (2019). Prediction of traffic volume by mining traffic sequences using travel time based PrefixSpan. *IET Intelligent Transport Systems*, vol. 13, no. 7, pp. 1199–1210. https://doi.org/10.1049/iet-its.2018.5165

4. Zhongjian, L, Xu, J, Zheng, K, Yin, H, Zhao, P & Zhou, X (2018). LC-RNN: A deep learning model for traffic speed prediction. *Proceedings of the Twenty-Seventh International Joint Conference on Artificial Intelligence (IJCAI-18)*, pp. 3470–3476.

5. Ganapathy, J & García Márquez, FP (2021). Data mining and information technology in transportation – a review. In: Xu, J, García Márquez, FP, Ali Hassan, MH, Duca, G, Hajiyev, A & Altiparmak, F (eds) *Proceedings of the Fifteenth International Conference on Management Science and Engineering Management. ICMSEM 2021. Lecture Notes on Data Engineering and Communications Technologies*, vol. 79. Springer, Cham. https://doi.org/10.1007/978-3-030-79206-0_64

6. Yang, Y, Cao, J, Qin, Y, Jia, L, Dong, H & Zhang, A (2018) Spatial correlation analysis of urban traffic state under a perspective of community detection. *International Journal of Modern Physics B*, vol. 32, no. 12, pp. 1–22.

7. Ganapathy, J (2023). Multi-criteria decision-making for sustainable transport: A case study on traffic flow prediction using spatial – temporal traffic sequence. In: García Márquez, FP & Lev, B (eds) *Sustainability: International Series in Operations Research & Management Science*, vol. 333. Springer, Cham. https://doi.org/10.1007/978-3-031-16620-4_9

8. Tang, K, Chen, S & Khattak, A (2018). A spatial-temporal multitask collaborative learning model for multistep traffic flow prediction. *Transportation Research Record: Journal of the Transportation Research Board*, vol. 2672, no. 45, pp. 1–13.

9. Shen, GC, Chen, Q, Pan, S, Shen, S & Liu, Z (2018). Research on traffic speed prediction by temporal clustering analysis and convolutional neural network with deformable kernels. *IEEE Access*, vol. 6, pp. 51756–51765.

10. Nguyen, D, Dow, C & Shiow-Fen, H (2018a). An efficient traffic congestion monitoring system on internet of vehicles. *Wireless Communications and Mobile Computing-Communication and Networking for Connected Vehicles*, no. 8, pp. 1–17.

11. Nguyen, H, Kieu, M, Wen, T & Cai, C (2018b). Deep learning methods in transportation domain: A review. *IET Intelligent Transport Systems*, vol. 12, no. 9, pp. 998–1004.

12. Ganapathy, J, García Márquez, FP & Ragavendra Prasad, M (2022). Routing vehicles on highways by augmenting traffic flow network: A review on speed up techniques. In: García Márquez, FP (eds) *International Conference on Intelligent Emerging Methods of Artificial Intelligence & Cloud Computing. IEMAICLOUD 2021. Smart Innovation, Systems and Technologies*, vol. 273. Springer, Cham. https://doi.org/10.1007/978-3-030-92905-3_11

13. Ermagun, A & Levinson, DM (2018). Development and application of the network weight matrix to predict traffic flow for congested and uncongested conditions. *Environment and Planning B: Urban Analytics and City*, vol. 6, no. 9, pp. 1684–1705.

14. Ganapathy, J (2021). Design of algorithm for IoT-based application: Case study on intelligent transport systems. In: García Márquez, FP & Lev, B (eds) *Internet of Things. International Series in Operations*

Research & Management Science, vol. 305. Springer, Cham. https://doi. org/10.1007/978-3-030-70478-0_11

15. Ganapathy, J & García Márquez, FP (2021). Travel time based traffic rerouting by augmenting traffic flow network with temporal and spatial relations for congestion management. In: Xu, J, García Márquez, FP, Ali Hassan, MH, Duca, G, Hajiyev, A & Altiparmak, F (eds) *Proceedings of the Fifteenth International Conference on Management Science and Engineering Management. ICMSEM 2021. Lecture Notes on Data Engineering and Communications Technologies,* vol. 78. Springer, Cham. https://doi. org/10.1007/978-3-030-79203-9_43

16. Ermagun, A & Levinson, D (2018) Spatio-temporal traffic forecasting: Review and proposed directions. *Transport Reviews,* vol. 38, no. 6, pp. 786–814.

17. Aakarsh, G, Aman, G, Samridh, S & Varun, S (2019). Factors affecting adoption of food delivery apps. *International Journal of Advanced Research,* vol. 7, no. 10, pp. 587–599.

18. Ana, IT, Javier, DS & Maitena, I (2018) Big data for transportation and mobility: recent advances, trends and challenges. *IET Intelligent Transport System,* vol. 12, no. 8, pp. 742–755.

19. Jayanthi, G & Ramya, M (2022). Transport data analytics with selection of tools and techniques for emergency medical services. In: *Smart Healthcare for Sustainable Urban Development.* IGI Global, Hershey, pp. 203–213.

20. Ryghaug, M, Subotički, I, Smeds, E, von Wirth, T, Scherrer, A, Foulds, C, Robison, R, Bertolini, L, Beyazit İnce, E, Brand, R, Cohen-Blankshtain, G, Dijk, M, Freudendal Pedersen, M, Gössling, S, Guzik, R, Kivimaa, P, Klöckner, C, Lazarova Nikolova, H, Lis, A, Marquet, O, Milakis, D, Mladenović, M, Mom, G, Mullen, C, Ortar, N, Paola, P, Oliveira, CS, Schwanen, T, Tuvikene, T & Wentland, A (2023). A social sciences and humanities research agenda for transport and mobility in Europe: Key themes and 100 research questions. *Transport Reviews.* https://doi.org/10.1080/01441647.2023.2167887

21. Cui, H, Meng, Q, Teng, TH & Yang, X (2023). Spatiotemporal correlation modelling for machine learning-based traffic state predictions: State-of-the-art and beyond. *Transport Reviews.* https://doi.org/10.1080/01441647.2023.2171151

22. Kiran, BR, et al. (2022, June). Deep reinforcement learning for autonomous driving: A survey. *IEEE Transactions on Intelligent Transportation Systems,* vol. 23, no. 6, pp. 4909–4926. https://doi.org/10.1109/TITS.2021.3054625

23. Nguyen, H, Kieu, L-M, Wen, T & Cai, C (2018). Deep learning methods in transportation domain: A review. *IET Intelligent Transport Systems,* vol. 12, pp. 998–1004. https://doi.org/10.1049/iet-its.2018.0064

24. Han, L & Huang, Y-S (2020). Short-term traffic flow prediction of road network based on deep learning. *IET Intelligent Transport Systems,* vol. 14, pp. 495–503. https://doi.org/10.1049/iet-its.2019.0133

25. Yuan, T, Da Rocha, W, Rothenberg, CE, Obraczka, K, Barakat, C & Turletti, T (2022). Machine learning for next-generation intelligent transportation systems: A survey. *Transactions on Emerging Telecommunications Technologies,* vol. 33, no. 4, p. e4427. https://doi.org/10.1002/ett.4427

Index

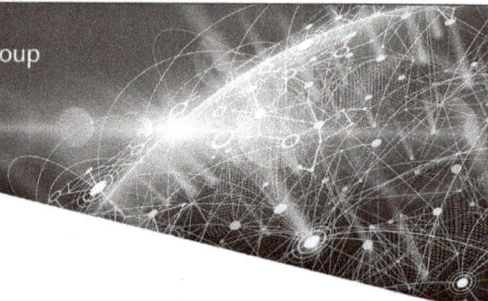

For Product Safety Concerns and Information please contact our EU
representative GPSR@taylorandfrancis.com
Taylor & Francis Verlag GmbH, Kaufingerstraße 24, 80331 München, Germany